Henry Lovejoy Ambler

Tin Foil and it's Combinations for Filling Teeth

Henry Lovejoy Ambler

Tin Foil and it's Combinations for Filling Teeth

ISBN/EAN: 9783744743969

Printed in Europe, USA, Canada, Australia, Japan

Cover: Foto ©berggeist007 / pixelio.de

More available books at **www.hansebooks.com**

TIN FOIL

COMBINATIONS FOR FILLING TEETH.

BY

HENRY L. AMBLER, M.S., D.D.S., M.D.,

Professor of Operative Dentistry and Dental Hygiene, in the Dental Department of Western Reserve University.
Member of the American Dental Association ; of the Ohio State Dental Society ; of the Northern Ohio Dental Association ; of the Cleveland City Dental Society.

THE S. S. WHITE DENTAL MFG. CO.,

LONDON :

CLAUDIUS ASH & SONS, Limited.

PREFACE.

BELIEVING that sufficient and well-deserved prominence was not being given to the use of tin foil and its combinations, the author decided to present a brief historical résumé of the subject, together with such practical information as he possesses, before the profession in order that it may have the satisfaction of saving more teeth, since that is the pre-eminent function of the modern dentist. One object is to meet the demand for information in regard to the properties and uses of tin foil; this information has been sought to be given in the simplest form consistent with scientific accuracy. The present use of tin is a case of the "survival of the fittest," because tin was used for filling teeth more than one hundred years ago. There is not a large amount of literature upon the subject, and no single text-book has treated the matter fully enough to answer the needs of both teacher and pupil. It is difficult for the student to collect and harmonize from the many different sources just the kind and amount of information required for his special use. Perhaps this work will be of assistance to scientific students and practical operators in the art of using tin foil, including all who wish in compact form an explanation of the facts and principles upon which the art is based. A good method to arouse in students an interest in the use of tin foil is to have them use it in operative technics, which is becoming an effective adjunct in every dental college. By this means a great factor will be brought to bear, and the result will be that hundreds of graduates

iii

every year will begin practice better qualified to save teeth
than if they had not known whatever may be learned about
this material. At the University of Pennsylvania, Department of Dentistry, session 1896-97, out of the total number
of fillings made in the clinical department (fractions omitted)
55 per cent. were gold, 15 per cent. tin, 10 per cent. amalgam.
This shows that tin has some very strong friends in the per-
sons of Professors Darby and Kirk.

The historical sketch of the development of the subject is
arranged in chronological order, and is given partly to show
that some old ideas and methods were good, and some
obviously incorrect when viewed in the light of more recent
developments. Part of the history will be new to the oldest
members of the profession, and the younger ones will certainly read it with interest. The work has been brought up
to date by considering all the properties and methods available. More names, good opinions, and dates could have
been given, but the writer believes that what is herein presented is enough to thoroughly substantiate his own opinions,
experiments, and practical applications. Some of the illus-
trations have been made especially for this work; the others
have been obtained through the courtesy of the owners.

"Let not the foggy doctrine of the superiority of gold in
all cases act on progress as the old medieval superstitions
acted on astronomy, physiology, zoology. Truth sought
after without misgiving, and the humblest as well as the
highest evidence taken in every case, and acted on with skill
and discrimination, will crown all with a high average of
success."

It is hoped that what has been said in this volume will
enable those who study it to save more teeth, and stimulate
them to make improvements on the material and methods,
doing much better than has been described or suggested.

CLEVELAND, OHIO, June, 1897.

"With soft and yielding lamina, and skill,
The practiced dental surgeon learns to fill
Each morbid cavity, by caries made,
With pliant tin; when thus the parts decayed
Are well supplied, corrosion, forced to yield
To conquering art the long-contested field,
Resigns its victim to the smiles of peace,
And all decay and irritation cease."

(Solyman Brown.)

The quantity of tin foil used measures the number of teeth saved with *metals* in any country during any historical period.

CONTENTS.

CHAPTER I.

CHAPTER II.

CHAPTER III.

CHAPTER IV.

CHAPTER V.

CHAPTER VI.

CHAPTER I.

MOSES, who was born 1600 B.C., mentions tin, and history records its use 500 B.C., but not for filling teeth; much later on, the Phœnicians took it from Cornwall, England, to Tyre and Sidon.

The alchemistic name for tin is Jove, and in the alchemistic nomenclature medicinal preparations made from it are called Jovial preparations.

Hindoo native doctors give tin salts for urinary affections. Monroe, Fothergill, and Richter claim to have expelled worms from the human system, by administering tin filings.

Blackie, in "Lays of Highlands and Islands," referring to tin as money, says,—

> "And is this all? And have I seen the whole
> Cathedral, chapel, nunnery, and graves?
> 'Tis scantly worth the tin, upon my soul."

"Tin-penny."—A customary duty formerly paid to the tithingmen in England for liberty to dig in the tin-mines.

In 1846, Tin (Stannum, symbol Sn) was found in the United States only at Jackson, N. H. Since then it has been found, to a limited extent, in West Virginia and adjoining parts of Ohio, North Carolina, Utah, and North Dakota. The richest tin mines of the world, however, are in Cornwall, England, which have been worked from the time of the Phœnician discovery.

The tin which is found in Malacca and Banca, India, is of great purity, and is called "Straits Tin" or "Stream Tin." It occurs in alluvial deposits in the form of small rounded grains, which are washed, stamped, mixed with slag and scoriæ, and smelted with charcoal, then run into basins, where the upper portion, after being removed, is known as the best refined tin. Stream tin is not pure metallic tin, but is the result of the disintegration of granitic and other rocks which contain veins of tinstone. Banca tin is 99.961 parts tin, 0.019 iron, 0.014 lead in 100 parts; it is sold in blocks of 40 and 120 pounds, and a bar 0.5 meter long, 0.1 broad, 0.005 deep can be bent seventy-four times without being broken. Subjected to friction, tin emits a characteristic odor.

Tin in solution is largely used in electro-metallurgy for plating. Pure tin may be obtained by dissolving commercial tin in hydrochloric acid, by which it is converted into stannous chlorid; after filtering, this solution is evaporated to a small bulk, and treated with nitric acid, which converts it into stannic oxid, which in turn is thoroughly washed and dried, then heated to redness in a crucible with charcoal, producing a button of tin which is found at the bottom of the crucible.

Pure tin may be precipitated in quadratic crystals by a slight galvanic current excited by immersing a plate of tin in a strong solution of stannous chlorid; water is carefully poured in so as not to disturb the layer of tin solution; the pure metal will be deposited on the plate of tin, at the point of junction of the water and metallic solution.

In the study of tin as a material for filling teeth, we have deemed it expedient to consider some of its physical characteristics, in order that what follows may be more clearly understood.

Tin possesses a crystallized structure, and can be obtained in well-formed crystals of the tetragonal or quadratic system (form right square prism), and on account of this crystalline structure, a bar of tin when bent emits a creaking sound, termed the "cry of tin;" the purer the tin the more marked the cry.

The specific gravity is 7.29; electrical state positive; fusing point 442° F.; tensile strength per square inch in tons, 2 to 3. Tensile strength is the resistance of the fibers or particles of a body to separation, so that the amount stated is the weight or power required to tear asunder a bar of pure tin having a cross-section of one square inch.

Tenacity: Iron is the most tenacious of metals. To pull asunder an iron wire 0.787 of a line in diameter requires a weight of 549 lbs. To pull asunder a gold wire of the same size, 150 lbs.; tin wire, 34 lbs.; gold being thus shown to be more than four times as tenacious as tin. (Fractions omitted.)

Malleability: Pure tin may be beaten into leaves one-fortieth of a millimeter thick, thus requiring 1020 to make an inch in thickness. Miller states that it can be beaten into leaves .008 of a millimeter thick, thus requiring 3175 to make an inch in thickness. Richardson says that ordinary tin foil is about 0.001 of an inch in thickness.

If the difficulty with which a mass of gold (the most malleable of metals) can be hammered or rolled into a thin sheet without being torn, be taken as one, then it will be four times as difficult to manipulate tin into thin sheets.

Ductility: If the difficulty with which gold (the most ductile of metals) can be drawn be taken as one, then it will be seven times as difficult to draw tin into a wire. At a temperature of 212° it has considerable ductility, and can be drawn into wire.

Among the metals, silver is the best conductor of

heat. If the conductivity of silver be taken as 100, then the conducting power of gold would be 53.2; tin, 14.5; gold being thus shown to be nearly four times as good a conductor of heat as tin. Among the metals, silver is the best conductor of electricity. If its electrical conductivity be taken at 100, then the conducting power of gold would be 77.96; tin, 12.36; gold being thus shown to be more than six times as good a conductor of electricity as tin.

Resistance to air: If exposed to dry, pure air, tin resists any change for a *great* length of time, but if exposed to air containing moisture, carbonic acid, etc., its time resistance is reduced, although even then it resists corrosion much better than copper or iron.

As to linear expansion, when raised from 32° to 212° F., aluminum expands the most of any of the metals. Taking its expansion as 1, that of tin would be 3, *i.e.*, aluminum expands three times as much as tin. (Dixon, "Vade Mecum.")

Solids generally expand equally in all directions, and on cooling return to their original shape. Within certain limits, metals expand uniformly in direct proportion to the increase in temperature, but the rate of expansion varies with different metals; thus, under like conditions, tin expands

nearly twice ($1\frac{3}{5}$) as much as gold, but the *rate* of
expansion for gold is nearly twice ($1\frac{7}{10}$) that of
tin.

The capacity for absorbing heat varies with each
metal; that of gold is about twice ($1\frac{3}{4}$) that of tin.

Tin has a scale hardness of about 4, on a scale of
12 where lead is taken as the softest and platinum
the hardest. (Dixon, "Vade Mecum.")

Tin has a scale hardness of about 2. (Dr. Miller.)   .

To fuse a tin wire one centimeter in diameter
requires a fusing current of electricity of 405.5
amperes. Up to 225° C., the rise in resistance to
the passage of an electric current is more rapid in
tin than in gold. In some minerals the current
follows the trend of the crystals.

Gold wire coated with tin, and held in the flame
of a Bunsen burner, will melt like a tin wire. At
1600° to 1800° tin boils and may be distilled.

CHAPTER II.

THE largest and most complete dental library in the world is owned by Dr. H. J. McKellops, of St. Louis. Upon his cheerful invitation, the writer visited that "Mecca," and through his kindness and assistance a complete search was made, which resulted in obtaining a great portion of the following historical facts with reference to the use of tin in dentistry:

"In 1783 I stopped a considerable decay in a large double under tooth, on the outside of the crown or near the gums, with fine tin foil, which lasted for a good number of years." ("A Practical and Domestic Treatise on Teeth and Gums," by Mr. Sigmond, Bath, England, 1825.)

"Fine tin foil or gold leaf may be injected into a cavity successfully, and retained securely for many years." (Joseph Fox, Dover, England, 1802.)

"The statement has been made several times that tin foil was used in the United States for filling teeth as early as 1800, at which time dentistry began to be cultivated particularly as a science and art, and was beginning to be regarded as of more importance than it formerly had been. The writer has not found any record of its use in this country

earlier than 1809. Tin may often be employed
with entire confidence. I have seen fillings forty-
one years old (made in 1809) and still perfect.
Several molars had four or five plugs in them,
which had been inserted at different periods during
the last half-century. I prefer strips cut from six
sheets laid upon each other. If the foil is well con-
nected, the cut edges will adhere firmly; if they do
not, the foil is not fit for use." (Dr. B. T. Whit-
ney, *Dental Register of the West*, 1850.) First
reference to the fact that tin is adhesive.

"Tin is desirable in all unexposed cavities. It
has a stronger affinity for acetic, citric, tartaric,
malic, lactic, and nitric acids than the tooth has: a
good material where the secretions are of an acid
character, it is better that the filling should waste
away than the tooth. One cavity in my mouth
was filled with gold, decay occurred, the filling was
removed; cavity filled with oxychlorid, which produced pain; filling removed; cavity filled with
gutta-percha, still experienced pain; filling removed; cavity filled with tin, and pain ceased in an
hour. A tin filling was shown in New York which
was sixty years old; made in 1811." (Dr. E. A.
Bogue, *British Journal of Dental Science*, 1871.)

"I have lately been removing tin pluggings (the
juices of the mouth having oxidated and dissolved

away the metal, so as to expose the teeth to decay) from teeth which I plugged fifteen years ago (1818) for the purpose of re-stopping with gold, and have in almost every instance found the bone of the tooth at the bottom of the pluggings perfectly sound and protected from decay." (J. R. Spooner, Montreal, 1833.)

In 1800 the number of dentists in the United States was about one hundred, and many of them were using tin foil for filling teeth.

In 1822 tin was employed by the best dentists, with hardly an exception; it grew in favor, especially for large cavities in molars, and for a cheaper class of operations than gold, but tin was not generally used until 1830. ("History of Dental and Oral Science in America.")

"Lead, tin, and silver corrode and become more injurious than the original disease, and will in every case ultimately prove the cause of destruction to the tooth, which might have been preserved by proper treatment." (Leonard Koecker, 1826, and "New System of Treating the Human Teeth," by J. Paterson Clark, London, 1829 and 1830.)

"Tin in situations out of reach of friction in mastication, as between two teeth, is like the tooth itself apt to be decomposed by acidity unless kept very clean." ("Practical and Familiar Treatise on

Teeth and Dentism," J. Paterson Clark, London, 1836.) Refer to what the same author said in 1829.

"Tin is used as a plugging material." ("The Anatomy, Physiology, and Diseases of the Teeth," by Thomas Bell, F.R.S., London, 1829.)

"Silver and tin foil, although bright when first put in a cavity, very soon change to a dark hue. resembling the decayed parts of the teeth which are of a bluish cast; besides this, they are not sufficiently pure to remain in an unchanged state, and frequently they assist in the destruction of a tooth instead of retarding it." ("Familiar Treatise on the Teeth," by Joseph Harris, London, 1830.)

"Tin is objectionable on account of rapid oxidation and being washed by the saliva into the stomach, as it may materially disorder it; the filling becomes so reduced that the cavity in which it has been inserted will no longer retain it, and acid fruits influence galvanic action." ("Every Man his Own Dentist," Joseph Scott, London, 1833.)

In 1836 Dr. Diaz, of Jamaica, used tin foil for filling teeth.

"Gold is now preferred, though tin, from its toughness when in the leaf, is perhaps the most suitable. Americans are superior to British in

FIG. 1.

FIG. 2.

filling." ("Plain Advice on Care of the Teeth," Dr. A. Cameron, Glasgow, 1838.)

"Tin foil is used for filling teeth." (S. Spooner, New York, 1838, "Guide to Sound Teeth.")

In 1838 Archibald Mc-Bride, of Pittsburg, Pa., used tin for filling cavities of decay.

The following facts were learned from Dr. Corydon Palmer: E. E. Smith, who had been a student of John and William Birkey, in Philadelphia, came to Warren, Ohio, in 1839, and among other things made the first gold plate in that part of the country. In operating on the anterior teeth, he first passed a separating file between them, excavated the cavity, and prepared the foil, *tin* or gold, in tapes which were cut transversely, every ·eighth of an inch, about three-quarters

of the way across.   Fig. 1 shows the size of tape
and the manner of cutting.   With an instrument
(Fig. 2) he drew the foil in from the labial surface,
using such portion of the tape as desired.

The instrument from which the illustration was
made was furnished by Dr. Palmer, and is shown
full size.   Instruments for use on posterior teeth
were short and strong, with as few curves as possible; no right and left cutters or pluggers were
used, and none of the latter were serrated, but had
straight, tapering round points, flat on the ends,
and of suitable size to fill a good portion of the
cavity.   He used what was termed Abbey's chemically pure tin foil, forcing it in hard, layer upon
layer,—as he expressed it, "smacked it up."   In
this manner he made tin fillings that lasted more
than thirty years.

In 1839 Dr. Corydon Palmer filled teeth with tin
foil, also lined cavities with gold and filled the remainder with tin.   In the same year he filled crown
(occlusal) cavities one-half full with tin and the
other half with gold, allowing both metals to come
to the surface, on the same plan that many proximal cavities are now filled.   (See Fig. 3, showing
about one-half of the cavity nearly completed with
tin cylinders.   The same plan was followed when
strips, or ropes, were used.)

"I filled cavities about two-thirds full with tin, and finished with gold." (S. S. Stringfellow, *American Journal of Dental Science*, 1839.)

"Tin foil is greatly used by some American dentists, but it is not much better than lead leaf." ("Surgical, Operative, and Mechanical Dentistry," L. Charles De Londe, London, 1840.)

"In 1841 there were about twelve hundred den-

FIG. 3.

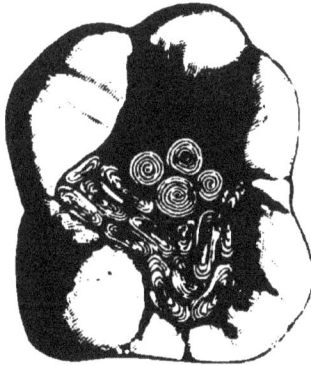

tists in the United States, many of whom were using tin, and there are circumstances under which it may be used not only with impunity, but advantage, but it is liable to change." (Harris.)

"I put in tin fillings, and at the end of thirty years they were badly worn, but there was no decay around the margins." (Dr. Neall, 1843.)

In 1843 Dr. Amos Westcott, of Syracuse, N. Y.,

filled the base of large cavities with tin, completing the operation with gold.

"Tin is used in the form of little balls, or tubes, but folds are better; introduce the metal gradually, taking care to pack it so that it will bear equally upon all points; the folds superimpose themselves one upon the other; thus we obtain a successive stratification much more exact and dense, and it is impossible there can be any void." ("Theory and Practice of Dental Surgery," J. Lefoulon, Paris, 1844.)

CHAPTER III.

"BESIDES gold, the only material which can be used with any hope of permanent success is tin foil. Some dentists call it *silver*, and a tooth which can-not be filled with it cannot be filled with anything else so as to stop decay and make it last very long. It can be used only in the back teeth, as its dark color renders it unsuitable for those in front. When the general health is good, and the teeth little predisposed to decay, this metal will preserve them as effectually perhaps as gold; but where the fluids of the mouth are much disordered it oxidizes rapidly, and instead of preserving the teeth rather increases their tendency to decay." (Dr. Robert Arthur, Baltimore, 1845, "A Popular Treatise on the Diseases of the Teeth.")

The false idea that a patient must have good health, normal oral fluids, and teeth little predisposed to decay, or else if filled with tin the decay would be hastened, originated with a German or English author, and has been handed down in works published since early in 1800. It even crept into American text-books as late as 1860, the authors of which now disbelieve it.

"Tin undergoes but little change in the mouth,

and may be used with comparative safety." ("Surgical, Mechanical, and Medical Treatment of the Teeth," James Robinson, London, 1846.)

"Tin is soft, and can be easily and compactly introduced, but it is more easily acted on by the secretions of the mouth than gold and is less durable, but in the mouth of a healthy person *it will last for years.* Still, inasmuch as it cannot be depended on in *all* cases, we are of the opinion that it should *never* be employed." ("The Human Teeth," James Fox, London, 1846.)

The italics are ours. Every metal has a limited sphere of usefulness, and it should not be expected that tin will contend single-handed against all the complicated conditions which caries presents.

"Of all the cheaper materials, I consider tin the best by far, and regard its use fully justifiable in deciduous teeth and in large cavities, as it is not every man who can afford the expense of nine leaves of gold and four hours of labor by a dentist on a single tooth." (Dr. Edward Taylor, *Dental Register of the West*, 1847.)

"I consider tin good for any cavity in a chalky tooth: it will save them better than anything else." (Dr. Holmes, 1848.)

"Tin can be used as a temporary filling, or as a matter of economy. It may be rendered imper-

vious to air and dampness, but it corrodes in most
mouths, unless it comes in contact with food in
chewing, and then it rapidly wears away; it does
not become hard by packing or under pressure, and
that it forms a kind of a union with the tooth is
ridiculous." (Dr. J. D. White, 1849, *Dental News
Letter*.)

"A tin plug will answer a very good purpose in
medium and large cavities for six years. Much
imposition has been practiced with it, and it is not
made as malleable as it should and can be. An
inferior article is manufactured which possesses
brilliancy and resembles silver. This is often
passed off for silver foil. No harm comes from
this deception except the loss of the amount paid
above the price for tin; but even this inferior tin
foil is better than silver." ("The Practical Family
Dentist," Dewitt C. Warner, New York, 1853.)

"Tin made into leaves is employed as a stopping
material; with sufficient experience it can be elab-
orated into the finest lines and cracks, and against
almost the weakest walls, and teeth are sometimes
lost with gold that might have been well preserved
with tin. I saw an effective tin stopping in a tooth
of Cramer's, the celebrated musical composer,
which had been placed there thirty-five years ago

by Talma, of Paris." ("The Odontalgist," by J. Paterson Clark, London, 1854.)

Refer to what the same author said in 1836.

"Tin is the best substitute for gold, and can often be used in badly shaped cavities where gold cannot." (Prof. Harris, 1854.)

"Tin is better than any mixture of metals for filling teeth." (Professor Tomes, London, 1859.)

In 1860 a writer said that "such a change may take place in the mouth as to destroy tin fillings which had been useful for years, and that tin was not entirely reliable in any case; it must not be used in a tooth where there is another metal, nor be put in the bottom of a cavity and covered with gold, for the tin will yield, and when fluids come in contact with the metals, chemical action is induced, and the tin is oxidized. Similar fillings in the same mouth may not save the teeth equally well. Filling is predicated on the nature of decay, for only on correct diagnosis can a proper filling-material be selected."

Reviewing the foregoing statement, we believe that a change may take place in the mouth which will destroy gold fillings (or ,the tooth-structure around them) much oftener than those of tin. It is now every-day practice to put tin into the same tooth with another metal; if the bottom of a cavity

is filled with tin properly packed, it will not yield
when completed with gold, and if the gold is tight,
the oral fluids cannot come in contact with both
metals and produce chemical action or oxidation;
similar fillings of gold in the same mouth do not
save the teeth equally well. Should we expect
more of tin in this respect, or discard it because it
is not always better than gold?

In Article V of the "New Departure Creed,"
Dr. Flagg says, "Skillful and scrupulous dentists
fill with tin covered with gold, thereby preventing
decay, pulpitis, death of the pulp, and abscess, and'
thus save the teeth."

In 1862 Mr. Hockley, of London, mentions tin
for filling, and the same year Dr. Zeitman, of Germany, recommended it as a substitute for gold, particularly for poor people.

"Is tin foil poisonous?   If not, why are our
brethren so reluctant to use it?   Is it nauseous?
If not, why not employ it?   Will it not preserve
the teeth when properly used?   Then why not
encourage the use of it?   Does its name signify
one too common in the eyes of the people, on
account of its daily use in the tin shops, or do patients murmur when the fee is announced, because
it is nothing but tin?   Is it not better than amalgam, although the patient may believe it less

costly? Eleven good plugs, twenty-nine years old, in one mouth demonstrates that tin will last as long as gold in many cases." (F. A. Brewer, *Dental Cosmos*, 1863.)

"So much tin foil is used for personal and domestic purposes that the following is important: Ordinary tin foil by chemical analysis contained 88.93 per cent. of lead; embossed foil, 76.57 per cent.; tea foil, 88.66 per cent.; that which was sold for the pure article, 34.62 per cent. Tin foil of above kind is made by inclosing an ingot of lead between two ingots of tin, and rolling them out into foil, thus having the tin on the outside of the lead." (Dr. J. H. Baldock, *Dental Cosmos*, 1867.)

The author used tin foil for filling the teeth of some of his fellow-students at the Ohio College of Dental Surgery in 1867.

"Amalgam should never be used in teeth which can be filled with tin, and most of them can be." (Dr. H. M. Brooker, Montreal, 1870.)

"I have used tin extensively, and found it more satisfactory than amalgam. Dentists ignore tin, because it is easier to use amalgam, less trouble. This is not right. If your preceptor has told you that amalgam is as good as tin, and he thinks so, let him write an article in its defense. Not one dentist in ten who has come into the profession within the last ten years knows how to make a tin filling, and only a few of the older ones know how to make a *good* one." (Dr. H. S. Chase, *Missouri Dental Journal*, 1870.)

"Among the best operators a more general use of tin would produce advantageous results, while among those whose operations in gold are not generally successful an almost exclusive use of tin would bring about a corresponding quantum of success to themselves and patients, as against repeated failures with gold. The same degree of endeavor which lacked success with gold, if applied to tin would produce good results and save teeth. A golden shower of ducats realized for gold finds enthusiastic admirers, but a dull gray shower for tin work is not so admirable, even though many of the teeth were no better for the gold as gold, nor so well off in the ultimate as with tin." (Dr. E. W. Foster, *Dental Cosmos*, 1873.)

In 1873 Dr. Royal Varney said, "I am heartily in favor of tin; it is too much neglected by our first-class operators."

"Tin stops the ends of the tubuli and interglobular spaces which are formed in the teeth of excessive vascular organization; if more teeth were filled with tin, and a smaller number with futile attempts with gold, people would be more benefited." (Dr. Castle, *Dental Cosmos*, 1873.)

"If cavities in teeth out of the mouth are well filled with tin, and put into ink for three days, no discoloration of the tooth (when split open) can be seen." (W. E. Driscoll, *Dental Cosmos*, 1874.)

"Tin makes an hermetical filling, and resists the disintegrating action of the fluids of the mouth. If an operator can preserve teeth for fifteen dollars with tin, which would cost fifty dollars with gold, ought he not to do so? Upon examination of the cavities from which oxidized plugs have been removed, these oxids will be found to have had a reflex effect upon the dentin; the walls and floors will be discolored and thoroughly indurated, and to a great degree devoid of sensitiveness, although they were sensitive when filled. Tin is valuable in case of youth, nervousness, impatience, high vitality of dentin, low calcification, and low pecuniosity." (Dr. H. Gerhart, *Pennsylvania Journal of Dental Science*, 1875.)

"Tin Foil for Filling Teeth." Essay by Dr. H. L. Ambler, read before the Ohio State Dental Society. (*Dental Register of the West*, 1875.)

"Some say that if tin is the material the cavity must be filled with, that it must be filled entirely

with it, but advanced teachings show differently."
(Dr. D. D. Smith, *Dental Cosmos,* October, 1878.)

"Frail teeth can be saved better with tin than
with gold. I never saw a devitalized pulp under a
tin filling." (Dr. Dixon, *Dental Cosmos,* May,
1880.)

"Tin may be used as a base for proximate fillings
in bicuspids or molars, in third molars, in children's
permanent molars, in the temporary teeth, and in
any cavity where the filling is not conspicuous."
(Dr. A. W. Harlan, *Independent Practitioner,* 1884.)

"Tin in blocks, mats, and tapes is used like non-
cohesive gold foil, but absence of cohesion prevents
the pieces from keeping their place as well as the
gold." ("American System of Dentistry," 1887.)

This is virtually saying that there is cohesion of
non-cohesive gold, and that for this reason it keeps
its place better than tin. It has always been supposed that there was no cohesion of layers of non-
cohesive gold, and as the tin is used on the non-
cohesive plan, therefore one keeps its place as well
as the other. We claim that generally in starting
a filling, tin will keep its place better than cohesive
or non-cohesive gold, because it combines some of
the cohesiveness of the former with the adaptability
of the latter.

"Tin will save teeth in many cases as well or

better than gold. Put a mat of tin at the cervical wall of proximate cavities in molars and bicuspids, and it makes a good filling which has a therapeutic effect on tooth-structure that prevents the recurrence of caries, probably because the infiltration of tin oxid into the tubuli is destructive to animal life. Where the filling is not exposed to mechanical force, there is no material under heavens which will preserve the teeth better." (Dr. Beach, *Dental Cosmos*, 1889.)

"I extracted a tooth in which I found a cavity of decay which had extended toward a tin filling, but stopped before reaching it; on examining the tooth-structure between the new cavity and the tin filling, it was found to be very hard, indicating apparently that there had been some action produced by the presence of the tin." (Dr. G. White, *Dental Cosmos*, 1889.)

"Pure tin in form of foil is used as a filling and also in connection with non-cohesive gold." (Mitchell's "Dental Chemistry," 1890.)

"Tin ranks next to gold as a filling-material." (Essig's "Dental Metallurgy," 1893.)

"Tin is good for children's teeth, when gold or amalgam is not indicated. It can be used in cavities which are so sensitive to thermal changes as to render the use of gold or amalgam unwise, but it can only be used in cavities with continuous walls, and should be introduced in the form of cylinders or ropes, with wedge-shaped pluggers having sharp deep serrations, thus depending upon the wedging or interdigitating process to hold the filling in the cavity." ("Operative Technics," Prof. T. E. Weeks, 1895.)

"Tin for filling teeth has been almost superseded by amalgam, although among the older practitioners (those who understand how to manipulate it) tin is considered one of the best, if not the very best metal known for preserving the teeth from caries. In consequence of its lack of the cohesive property, it is introduced and retained in a cavity upon the wedging principle, the last piece serving as a keystone or anchor to the whole filling. Each piece should fill a portion of the cavity from the bottom to the top, with sufficient tin protruding from the cavity to serve for thorough condensation of the surface, and the last piece inserted should have a retaining cavity to hold it firmly in place. The foil is prepared by folding a whole or half-sheet and twisting it into a rope, which is then cut into suitable lengths for the cavity to be filled." (Frank Abbott, "Dental Pathology and Practice," 1896.)

"Forty-three years ago, for a young lady fourteen years of age, I filled with non-cohesive gold all the teeth worth filling with this metal; the rest I filled with tin. Three years after that there was not a perfect gold filling among the whole number, and yet the tin fillings were just as good as when made. The explanation as to why the tin fillings lasted so much longer that the gold ones was, that there must have been something in the tin that had an affinity for the teeth and the elements that formed the dentin, by which some compound was formed, or else it must have been in the adaptation." (Dr. H. Gerhart, *Dental Cosmos*, January, 1897.)

CHAPTER IV.

At the World's Columbian Dental Congress, held in Chicago, August, 1893, the author presented an essay on "Tin Foil for Filling Teeth."

During the discussion of the subject, the following opinions were elicited:

Dr. E. T. Darby: "I have always said that tin was one of the best filling-materials·we have, and believe more teeth could be saved with it than with gold. I have restored a whole crown with tin, in order to show its cohesive properties; the essayist has paid a very high and worthy tribute to tin."

Dr. R. R. Freeman: "I have used tin foil for twenty-five years, and know that it has therapeutic properties, and is one of the best filling-materials, not excepting gold."

Madam Tiburtius-Hirschfield: "I heartily indorse the use of tin, and have tested its cohesive properties by building up crowns."

Dr. A. H. Brockway: "I am a strong believer in the use of tin, on account of its adaptability, and the facility with which saving fillings can be made with it."

Dr. Gordon White: "After having used tin for

nine years, I claim that it is the best filling-material that has been given to our profession."

Dr. C. S. Stockton: "Tin is one of the best materials for saving teeth, and we should use it more than we do."

Dr. James Truman: "I use tin strictly upon the cohesive principle, and would place it in all teeth except the anterior ones, but would not hesitate to fill these when of a chalky character."

Dr. Corydon Palmer: "For fifty-four years I have been a firm advocate of the use of tin, and I have a filling in one of my teeth which is forty years old."

Dr. William Jarvie: "I rarely fill a cavity with gold for children under twelve years of age that I want to keep permanently, but use tin, and in five or ten years, more or less, it wears out. Still, it can easily be renewed, or if all the tin is removed we find the dentin hard and firm. The dentist is not always doing the best for his patients if he does not practice in this way."

Dr. C. E. Francis: "I have proved positively that tin foil in good condition is cohesive, and my views have been corroborated by dentists and chemists."

Dr. James E. Garretson: "Tin foil is cohesive,

and can be used the same as gold foil, and to an extent answers the same purpose."

Dr. C. R. Butler: "Tin is cohesive and makes a first-class saving filling."

Dr. W. C. Barrett: "Tin is as cohesive as gold, and if everything was blotted out of existence with which teeth could be filled, except tin, more teeth would be saved." '

Dr. L. D. Shepard: "Tin possesses some antiseptic properties for the preservation of teeth that gold does not."

Dr. W. D. Miller: "I use tin foil in cylinders, strips, and ropes, on the non-cohesive plan, but admit that it possesses a slight degree of cohesiveness, and when necessary can be built up like cohesive gold by using deeply serrated pluggers."

Dr. Benjamin Lord says, "It is said that we know the world, or learn the world, by comparison. If we compare tin foil with gold foil, we find that the tin, being softer, works more kindly, and can be more readily and with more certainty adapted to the walls, the inequalities, and the corners of the cavities.

"We find also that tin welds—mechanically, of course—more surely than soft gold, owing to its greater softness; the folds can be interlaced or forced into each other, and united with more cer-

tainty, and with so much security that, after the
packing and condensing are finished, the mass may
be cut like molten metal.

"I contend moreover that for contouring the
filling or restoring the natural shape of the teeth,
where there are three walls remaining to the cavity,
tin is fully equal to gold, and in some respects even
superior; as tin can be secured, where there is very
little to hold or retain the filling, better than gold,
owing to the ease and greater certainty of its adaptation to the retaining points or edges of the cavity.

"It will be said, however, that tin fillings will
wear away.   The surfaces that are exposed to mastication undoubtedly will wear in time; but the filling does not become leaky if it has been properly
packed and condensed, nor will the margins of the
cavity be attacked by further decay on that account.

"Altogether, I believe that we can make more
perfect fillings with tin than we can with gold, taking all classes of cavities; but it must not be understood that it is proposed that tin should ever take
the place of gold where the circumstances and conditions indicate that the latter should be used.   Of
course, the virtue is not in the gold or the tin, but
in the mechanical perfection of the operation, and
tin having more plasticity than gold, that perfection can be secured with more ease and certainty.

"If we compare tin with amalgam, we must certainly decide in favor of the former and give it preference; as if it is packed and condensed as perfectly as may be, we know just what such fillings will do every time. We know that there will be no changes or leakage of the fillings at the margins; whereas, with amalgam, the rule is shrinkage of the mass, and consequently the admission of moisture around the filling, the result being further decay. It is not contended that this is always the result with amalgam, but it is the general rule; yet we must use amalgam, as there are not a few cases where it is the best that we can do; but it is to be hoped; and I think it may be said, that as manipulative skill advances, amalgam will be less and less used.   For so-called temporary work, very often I prefer tin to gutta-percha, as it makes a much more reliable edge and lasts longer, even when placed and packed without great care."—*N. Y. Odon. Society Proceedings*, page 51, 1894.

One of the main reasons which induced the writer to begin the use of tin foil *(Stannum Foliatum)* for filling teeth, in 1867, was the fact that amalgam filling failures were being presented daily.   Believing that tin could do no worse, but probably would do better, we banished amalgam from the office for the succeeding seven years, using in the place of it tin, oxychlorid, and gutta-percha. Since that time we have seen no good reason for abandoning the use of tin, as time has proved it worthy of great confidence. There is no better dental litmus to distinguish the conservative from the progressive dentist.

If we take a retrospective view and consider what tin foil was thirty years ago, we do not wonder that so many operators failed to make tight, good-wearing fillings. As it came from the manufacturer it looked fairly bright, but after being exposed to the air for a short time it assumed a light brassy color, and lost what small amount of integrity it originally possessed. This tin was not properly refined before beating, or something was put on the foil while beating, so that it did not have the clean, bright surface and cohesive quality which our best foil now has. No. 4 was commonly used, but it would cut and crumble in the most provoking manner. Fillings were made by using mats, cylinders, tapes, and ropes, with hand-pressure, on the plan for manipulating non-cohesive gold foil, but it was difficult to insert a respectable approximal filling.

From the best information obtainable, the writer believes that Marcus Bull (the predecessor of Abbey) was the first to manufacture and sell tin foil in the United States, as he began the manufacture of gold foil at Hartford in 1812.

Several years ago a radical change came about in the preparation of tin foil, for which the manufacturer should have his share of the credit, even if the dentist did ask for something better, for the quality depends largely upon the kind and condition of the tin used and on the method of manufacture.

For making tin foil for filling teeth, the purest Banca tin that can be obtained is used. The tin is melted in a crucible under a cover of powdered charcoal. It is then cast into a bar and rolled to the desired thickness, so that if No. 6 foil is to be made, a piece one and one-half (1½) inches square would weigh nine grains. This ribbon is then cut into lengths of about four feet, and spread on a smooth board slanted, so that the end rests in a vat of clean water.

Then apply to the exposed surface of the ribbon diluted muriatic acid, and immediately wash with a strong solution of ammonia. Turn the ribbon and treat the other side in the same way. It is then washed and rubbed dry. The object of using the acid is to remove stains and whiten the tin, and the ammonia is used to neutralize the effect of the acid.

The strips are then cut into pieces one and a half

4

inch square, filled into a cutch and beaten to about
three inches square. It is then removed from the
cutch and filled into a mold, and further beaten to
the desired size. When the ragged edges are
trimmed off, the foil is ready for booking.

It takes skill and experience to beat tin foil, for it
is not nearly as malleable as gold; up to No. 20 it
is usually beaten, but higher numbers are prepared
by rolling. In each case the process is similar to
that employed in preparing gold foil. The number on the book is supposed to indicate the weight
or thickness of the leaf. On the lower numbers
the paper of the book leaves its impression.

On weighing sheets of tin foil from different
manufacturers a remarkable discrepancy was found
between the number on the book and the number
of grains in a sheet, viz: Nos. 3, 4, 5, weighed 7 gr.
each; No. 6, 9 gr.; No. 8, from 9 to 18 gr.; No. 10,
from 14 to 15 gr.; No. 20, 18 gr. In some instances
the sheets in the same book varied three grains.
We submit that it would be largely to the advantage of both manufacturer and dentist to have the
number and the grains correspond. No dentist
wishes to purchase No. 8 and find that he has No.
18; no one could sell gold foil under like circumstances. Of the different makes tested, White's
came the nearest to being correct. The extra

tough foil which can now be obtained is chemically pure, and with it we can begin at the base of any cavity, and with mallet or hand force produce a filling which will be one compact mass, so that it can be cut and filed; yet in finishing, it will not bear so severe treatment as cohesive gold. Always handle tin foil with clean pliers, never with the fingers; and prepare only what is needed for each case, keeping the remainder in the book placed in the envelope in which it is sold, otherwise extraneous matter collects upon it, and it will oxidize *slightly* when exposed to the air for a *great* length of time.

Before using tin foil, a few prefer to thoroughly crumple it in the hands or napkin, under the impression that they thus make it more pliable and easier to manipulate.

A piece of blue litmus paper moistened and moved over a sheet of tin foil will occasionally give an acid reaction, probably owing to the acid with which it is cleaned before beating not having been thoroughly removed. Foil held under the surface of distilled water and boiled for five minutes, then left until the water is cold, removed and dried, shows it has been annealed, which makes it work easily, but not as hard a filling can be made from it as before boiling.

In selecting and using this material for filling, we are able fully to protect the cavity; and if we understand the material, and how to manipulate it, we will surely succeed. This statement demands serious attention, and appeals to every one who is anxious to practice for the best interests of his patients; then let us make a thorough study of the merits of the method and material.

Until recently, the term cohesion had but one special meaning to dentists, and that as applied to gold for filling teeth; being understood as the property by which layers of this metal could be united without force so as to be inseparable. The writer claims that good tin foil in proper condition is cohesive when force is applied, and can be used for filling teeth in the same manner in which cohesive gold foil is used. This claim has been confirmed by several dentists, as noted in another part of this volume.

"Cohesion is the power to resist separation, and it acts at insensible distances. The integral particles of a body are held together by cohesion, the constituent parts are united by affinity.

The attraction between atoms of pure tin represents cohesion. Marble is composed of lime and carbonic acid, which are united and held together by affinity.

The condition which obtains in the tin may be called cohesion, adhesion, welding, or interdigitation, but the fact remains that layers of tin foil can be driven together into a solid mass, making a tight filling with less malleting than is required for gold; if it is overmalleted, the receiving surface is injured.

On account of its pliability it is easily adapted to the walls and margins, and a perfect fit is made, thus preventing capillary action and preventing further caries. Of all the metals used for filling it is the best tooth-preserver and the most compatible with tooth-substance, and the facility with which a saving filling can be made largely commends it.

Tin has great possibilities, and has already gained a high position as a filling-material. Upon the knowledge we possess of the possibilities and limitations of tin as a filling-material, and our ability to apply that knowledge, will largely depend our success in preserving teeth.

It is a good material for filling many cavities in the temporary teeth, and children will bear having it used, because it can be placed quickly, and but little force is required to condense one or two layers of No. 10 foil. The dentin in young teeth has a large proportion of organic material, for which

reason, if caries takes place, many believe it is
hastened by thermal changes. Gold fillings in
such teeth might prevent complete calcification, on
account of the gold being so good a conductor; but
if tin is used, there is much more probability of calcification taking place, because of its low conductivity and its therapeutic influence. It does not
change its shape after being packed into a cavity.
Under tin, teeth are calcified and saved by the
deposit of lime-salts from the contents of the dentinal tubuli. This is termed progressive calcification.

Like other organs of the human body, the teeth
are more or less subject to constitutional change.
The condition in which we find tooth-structure
which needs repairing or restoring should be a sure
indicator to us in choosing a filling-material. Up
to the age of fourteen, and sometimes later, we find
many teeth which are quite chalky. In some
mouths also, at this period, the fluids are in such a
condition that oxychlorid and oxyphosphate do not
last long; for some reason amalgam soon fails,
while gutta-percha is quickly worn out on an occlusal surface. In all such cases we recommend
tin, even in the anterior teeth, for as the patient
advances in years the tooth-structure usually becomes more dense, so that, if desirable, the fillings

can be removed, and good saving operations can be made with gold. By treating cases in this manner very little, if any, tooth-structure is lost.

The teeth of the inhabitants of Mexico and Guatemala are characteristic of their nervous and nervo-lymphatic temperaments; children ten years of age often have twenty-eight permanent teeth, and they are generally soft or chalky, but our dentists there report good success in saving them with tin.

In filling this class of teeth, we should be very careful not to use force enough to injure the cavity-margin, for if this occurs, a leaky filling will probably be the result. Still, we have seen some cases where *slight* imperfections at the margin, which occurred at the time of the filling or afterward, did no harm, because the deposit of tin oxid filled up the ends of the tubuli, thus preventing caries. We believe that this bar to the progress of caries is set up more frequently when tin is used than with any other metal under like conditions.

CHAPTER V.

In some mouths tin does not discolor, but retains a clean, unpolished tin color, yet when there is a sesquioxid of the metal formed, fillings present a grayish appearance. In the same mouth some fillings will be discolored, while others are not. As a general rule, proximal fillings are most liable to show discoloration. Perhaps one reason is that on occlusal and buccal surfaces they are subject to more friction from mastication, movements of the cheeks, and the use of the brush.

We have seen a large number of fillings which were not blackened, yet were saving the teeth perfectly, thus proving to a certainty that blackening of tin in the tooth-cavity is not absolutely essential in order to obtain its salvatory effects as a filling-material.

Where there is considerable decomposition of food which produces sulfuretted hydrogen, the sulfid of tin may be formed on and around the fillings; it is of a yellowish or brownish color, and as an antiseptic is in such cases desirable. To offset the discoloration, we find that the sulfid is insoluble, and fills the ends of the tubuli, thus lending its aid in preventing further caries. A sulfid is a com-

bination of sulfur with a metal or other body. A
tin solution acted on by sulfuretted hydrogen
(H_2S) produces a dark-brown precipitate (SnS),
stannous salt, which is soluble in ammonium sulfid
$(NH_4)_2 S_2$; this being precipitated, gives (SNS_2)
stannic salt, which is yellow. Brown precipitates
are formed by both hydrogen sulfid and ammonium
sulfid, in stannous solutions. Yellow precipitates
are formed by hydrogen sulfid and ammonium /
sulfid in stannic solutions. The yellow shade is
very seldom seen on tin fillings; the dark brown is
more common.

An oxid is a combination of oxygen with a metal
or base destitute of an acid. In oxidation the oxygen that enters into combination is not sufficient to
form an acid. The protoxid of tin (SnO) is black,
and can be obtained from chlorid of tin, or by *long*
exposure of tin to the atmosphere. The oxygen in
the saliva helps to blacken the tin, and the metallic
oxid penetrates the dentin more or less, acting as
a protection, because it is insoluble. Oxygen is
the only element which forms compounds with all
others, and is the type of electro-negative bodies;
it combines with all metals, therefore with tin, and
in many cases only the metal is discolored, and not
the tooth. Steam boilers are made tight by oxidation.

Where there is complete oxidation, the tooth is blackened to but a very slight depth, and the oxid fills the ends of the tubuli, thus affording an additional barrier to the entrance of caries. The filling itself will prevent caries, but oxidation acts as an assistant.

"In the mouth, a suboxid is more likely to be formed than a protoxid, but both are black; sulfur and oxygen are capable of acting on tin under favorable circumstances, such as warmth, moisture, full contact, condensation of elements, and their nascent conditions; the first three are always present in the mouth. The protosulfuret of tin is black." (Dr. George Watt.) Others give the color as bluish-gray, nearly black.

Experiments show that slight galvanic currents exist between fillings of dissimilar metals in the mouth, and practical experience demonstrates that these currents occasionally produce serious results.

Direct galvanic currents do not decompose normal teeth by true electrolysis, but acids resulting from decomposition of food and fluids react upon the lime constituents of the teeth and promote secondary caries.

When two metals are so situated in the mouth that the mucous membrane forms a connecting conductor and the fluids are capable of acting on

one metal, galvanic action is established sufficient
to decompose any of the binary compounds contained in these fluids; the liberated nitrogen and
hydrogen form ammonia, which being exposed to
the action of oxygen is decomposed and nitric oxid
formed, resulting in nitric acid. We also have in
the mouth air, moisture, and decomposing nitro-
genous food to assist in the production of nitric
acid.

"Galvanic action is more likely to develop hydrochloric acid, for the chlorids of sodium and potas-
sium are present in the normal saliva and mucus,
and when decomposed their chlorin unites with the
hydrogen derived from the water of the saliva."
(Dr. George Watt.)

The fact should also be noted that both nitric and
hydrochloric acids are administered as medicine,
and often assist in producing decay.

When there is a battery formed in a mouth containing tin fillings and gold fillings, and the fluids
of the mouth are the exciting media, tin will be
the positive element and gold the negative element;
thus when they form the voltaic pair, the tin becomes coated or oxidized and the current practically ceases.

There is more or less therapeutical and chemical
action in cavities filled with tin, and its compati-

bility and prophylactic behavior as a filling-material depends partly upon the chemical action which occurs.

Some dentists fill sensitive cavities with tin, in order to secure gentle galvanic action, which they believe to be therapeutic, solidifying the tooth-structure.

"Tin possesses antiseptic properties which do not pertain to gold for arresting decay in frail teeth; it not only arrests caries mechanically, but in chalky (imperfect) structure acts as an antacid element in arresting the galvanic current set up between the tooth-structure and filling-material." (Dr. S. B. Palmer.) If the metal is acted on, the tooth is comparatively safe; if the reverse, it is more or less destroyed. The galvanic taste can be produced by placing a piece of silver on the tongue and a steel pen or piece of zinc under it; then bring the edges of the two pieces together for a short time, rinse the saliva around in the mouth, and the peculiar flavor will be detected.

"In 1820 attention was called to the injurious effects of the galvanic current on the teeth, and dentists were advised never to use tin and amalgam in the same mouth.

"A constant galvanic action is kept up in the mouth when more than one kind of metal is used in filling teeth, and galvanism is often the cause of extensive injury to the teeth. The most remarkable case I ever saw was that of a lady for whom I filled several teeth with tin. After a time decay took place around some of the fillings. I removed them and began to refill, but there was so much pain I could not proceed. I found that by holding a steel plugger an inch from the tooth I could give her a violent galvanic shock. I observed that the exhalation of the breath increased the evolution of galvanism." (Dr. L. Mackall, *American Journal of Dental Science*, 1839.)

"When a faulty tooth in the upper jaw had been stopped from its side with tin, the interstice between it and the adjoining tooth being quite inconsiderable, while the upper surface of a tooth not immediately beneath it in the lower jaw was stopped with the same metal, I have known a galvanic shock regularly communicated from one tooth to the other when by the movement of jaws or cheeks they were brought near together." (Dr. E. Parmly, *American Journal of Dental Science*, 1839.)

"An interesting debate here sprung up on the action where two metals are used in one filling, such as gold and tin, the saliva acting as a medium, and where the baser metal is oxidized by exhalents and by imbibition through the bony tooth-structure." (Pennsylvania Society of Dental Surgeons, 1848.)

"A patient came to me and complained of pain in the teeth. Upon examination I found an amal gam filling next to one of tin. With a file I made a V-shaped separation, when they experienced immediate relief from pain." (Dr. Nevill, *American Journal of Dental Science*, 1867.)

In regard to the decay of teeth being dependent on galvanic action present in the mouth, Dr. Chase, in 1880, claimed that a tooth filled with gold would necessarily become carious again at the margin of the cavity, wherever the acid secretions constantly bathe the filling and tooth-substance. A tooth filled with amalgam succumbs to this electro-chemical process less rapidly, while one filled with tin still longer escapes destruction. The comparative rapidity with which teeth filled with gold, amalgam, or tin, are destroyed is expressed by the numbers 100, 67, 50. He prepared pieces of ivory of equal shape and size, bored a hole in each, and filled them. After they had been exposed to the action of an acid for one week, they had decreased in weight,— viz, piece filled with gold, 0.06; amalgam, 0.04; tin, 0.03.

"With tin and gold, some have the superstition

that the electricity attendant upon such a filling will
in some way be injurious to the tooth; it matters
not which is on the outside, when rolled and used as
non-cohesive cylinders each appears. We say that
neither experimentally, theoretically, nor practically can any good or bad result be expected from
the electrical action of a tin-gold filling on tooth-
bone, and neither will the pulp be disturbed."
(Dr. W. D. Miller, *Independent Practitioner*, August,
1884.)

"When the bottom of a cavity is filled with tin
which is tightly (completely) covered with gold,
there is *practically* no galvanic action and there is
no current generated by contact of tin and gold,—
i.e., no current leaves the filling to affect the dentin.
That portion of tin which forms the base is more
positive than a full tin filling would be. The effect
is to cause the surface exposed to dentin to oxidize
more than tin would do alone; in that there is a
benefit. In very porous dentin there is enough
moisture to oxidize the tin, by reason of the current
set up by the gold." (Dr. S. B. Palmer.)

Electricity generated by heat is called thermo-
electricity. If a cavity with continuous walls is
half filled with tin and completed with gold, or half
filled with silver and completed with gold, and the
junctions of the metal are at $20\frac{1}{2}°$ C. and $19\frac{1}{2}°$ C.,

if the electrical action between the tin and gold be
1.1, the action between the silver and gold will be
1.8, thus showing the action in silver and gold to be
nearly two-thirds more than in the tin and gold, a
deduction which favors the tin and gold.

Rubbing two different substances together is a
common method of producing an electric charge.
Is there not more electricity generated during mastication on metal fillings than when the jaws are at
rest? Friction brings into close contact numerous
particles of two bodies, and perhaps the electrical
action going on more or less all the time through
gold fillings (especially when other metals are in
the mouth) accounts for a powdered condition of
the dentin which is sometimes found under cohesive gold fillings, but not under tin.

CHAPTER VI.

WHITE caries, the most formidable variety known, may be produced by nitric acid, and in these cases all the components of the tooth are acted upon and disintegrated as far as the action extends. In proximal cavities attacked by this kind of caries, separate freely on the lingual side, and fill with tin. When such fillings have been removed the dentin has been found somewhat discolored and greatly solidified as compared to its former condition; this solidification or calcification is more frequent under tin than gold, which is partly due to the tin as a poor conductor of heat. Nature will not restore the lost part, but will do the next best thing—solidify the dentin. In some cases, under tin, the pulp gradually recedes, and the pulp-cavity is obliterated by secondary dentin. In other cases the pulps had partly calcified under tin. It has been known for years that tin would be tolerated in large cavities very near the pulp without causing any trouble, and one reason for this is its low conducting power. Attention is called to the fact that gold is nearly four times as good a conductor of heat as tin, and more than six times as good a conductor of electricity. Where tin fill-

5

ings are subject to a large amount of attrition, they
wear away sooner or later, but this is not such a
great detriment, for they can easily be repaired or
replaced, and owing to the concave form produced
by wear the patient is liable to know when a large
amount has been worn away. That portion against
the wall of the cavity is the last removed by wear,
so that further caries is prevented so long as there
is any reasonable amount of tin left. If at this time
the tooth has become sufficiently solidified, proper
anchorage can be cut in the tin or tooth, one or
both, as judgment dictates, and the filling completed with gold. A tin filling, confined by four
rather frail walls, may condense upon itself, but it is
so soft and adaptable that the force which condenses it continually secures the readaptation at
the margin; thus there will be no leakage or caries
for years. Owing to its softness and pliability, it
may be driven into or onto the tubuli to completely
close them from outside moisture, and with a hand
burnisher the tin can be made to take such a hold
on dry, rough tubuli that a cutting instrument is
necessary to remove all traces of it.

Tin foil has been found in the market that under
a magnifying glass showed innumerable tiny black
specks, which, upon being touched with an instrument, crumbled away, leaving a hole through the

foil. More than likely, some of the failures can be attributed to the use of such foil. Good tough foil, well condensed by hand or mallet force, stays against the walls of a cavity and makes a tight filling, and ought to be called as near perfect as any filling, because it preserves the tooth, and gives a surface which will wear from five to twenty years, depending upon the size and location of the cavity and tooth-structure. Buccal cavities in the first permanent molars, and lingual cavities in the superior incisors, filled for children from six to eight years of age, are still in good condition after a period of twenty years. Perhaps the limit is reached in the following cases, all in the mouths of *dentists:* One filling forty years old; one forty-two; four on the occlusal surface, fifty; in the latter case gold had been used in other cavities and had failed several times. Lingual cavities in molars and bicuspids can be perfectly preserved with tin. Tapes of No. 10 foil, from one to three thicknesses, can be welded together and will cohere as well or better than semi-cohesive gold foil, and it can be manipulated more rapidly; therefore, if desirable, any degree of contour can be produced, but the contour will not have the hardness or strength of gold, so in many cases it would not be practicable to make extensive contours with tin, owing to its physical characteristics.

No. 10 will answer for all cases, and it is not as liable to be torn or cut by the plugger as a lower number, but one need not be restricted to it, as good fillings can be made with Nos. 4, 6, or 8. More teeth can be saved with tin than with any other metal or metals, and the average dentist will do .better with tin than with gold.   It is invaluable when the patient is limited for time or means, and also for filling the first permanent molars, where we so often find poor calcification of tooth-structure. In cases of orthodontia, where caries has attacked a large number of teeth, it is well to fill with tin, and await further developments as to irregularity and caries.

If cavities are of a good general retaining form, that will be sufficient to hold the filling in place; but if not, then cut slight opposing angles, grooves, or pits.   Cavities are generally prepared the same as for gold, except where there is a great deal of force brought upon the filling; then the grooves or pits may be a little larger; still, many cavities can be well filled with less excavating than required for gold, and proximal cavities in bicuspids and molars, where there is sufficient space, can be filled without removing the occlusal surface, and here especially should the cavities be cut square into the teeth, so as not to leave a feather edge of tin when the filling is finished, as that would invite further caries and prove an obstruction to cleansing the filling with floss.

In proximal cavities involving the occlusal surface, cut the cervical portion down to a strong square base, with a slight pit, undercut, or angle, at the buccal and lingual corners; where there is sufficient material, a slight groove across the base, far enough from the margin so that it will not be broken out, can be made in place of the pit, undercut, or angle; then cut a groove in the buccal and lingual side (one or both, according to the amount of material there is to work upon), extending from the base to the occlusal surface; in most of these cases the occlusal grooves or pits would have to be excavated on account of caries; thus there would be additional opportunity for anchorage. In place of the grooves the cavity may be of the dovetail form. In nearly all proximal cavities in bicuspids and molars, some form of metal shield, or matrix, is of great advantage, as they prevent the tin from crushing or sliding out. By driving the tin firmly against the metal, a well-condensed surface is secured; and as the metal yields a little, we can with a bevel or thin plugger force the tin slightly between the metal and the margin of the cavity, thus making sure of a tight filling, with plenty of material to finish well. After removing the metal, condense with thin burnishers and complete the finish the same as for gold. Where no shield or matrix is used, or where it is used and removed before completing the filling, it is often desirable to trim the cervical border, for in either case there is more light and room to work when only a portion of the cavity has been filled. Tin cuts so much easier than gold, it is more readily trimmed down level with all cervical margins.

Be sure that all margins are made perfect as the work progresses, and if the cavity is deep and a wide shield shuts out the light, then use a narrow one, which can be moved toward the occlusal surface from time to time.

In filling the anterior teeth when the labial wall is gone, and the lingual wall intact or nearly so, use a piece of thin metal three-quarters of an inch long and wide enough to cover the cavity in the tooth to be filled, insert it between the teeth, and bend the lingual end over the cavity; the labial end is bent out of the way over the labial surface of the adjoining tooth, as shown in Fig. 4. When the labial wall is intact or nearly so, access to the cavity should be obtained from the lingual side, and in

this case the bending of the shield would be re-
versed, as shown in Fig. 5.   The shield is not absolutely essential, but it helps support the tin, and
also keeps a separation.

It is preferable to save the labial wall and line it
with (say) five layers of No. 4 semi-cohesive gold

FIG. 4.   FIG 5.

folded into a mat and extended to the outer edge
of the cavity; this gives the tooth a lighter shade,
and bicuspids or molars can be filled in the same
manner.   Cases are on record where incisors with
translucent labial walls, filled by this method, have
lasted from twenty-three to thirty-seven years.

CHAPTER VII.

FOR the last ten years the writer has been using tin at the cervical margin of proximal cavities in bicuspids and molars, especially in deep cavities (now an accepted practice), and he finds that it prevents further caries oftener than any other metal or combination of metals he has ever seen used. In filling such cavities, adjust the rubber, and use a shield or matrix of such form as to just pass beyond the cervical margin; this will generally push the rubber out of the cavity, but if it does not, then form a wedge of wood and force between the metal and the adjoining tooth, thus bringing the metal against the cervical margin, and if a small film of rubber should still remain in the cavity, it may be forced out by using any flat burnisher which will reach it, or it can be dissolved out with a little chloroform. Fill from one-fourth to one-half of the cavity with tin, and complete the remainder with gold when the tooth is of good structure; this gives all the advantages of gold for an occlusal surface.

Before beginning with the gold, have the tin solid and square across the cavity, and the rest of the cavity a good retaining form, the same as for gold filling; then begin with a strip of gold slightly annealed and mallet it into the tin, but do not place too great reliance upon the connection of the metals to keep the filling in place.

On the same plan, proximal cavities in the anterior teeth can be filled, and also buccal cavities in molars, especially where they extend to the occlusal surface. The cervical margin should be well covered with tin thoroughly condensed, thus securing perfect adaptation, and a solid base for the gold with which the filling is to be completed. Time has fully demonstrated that the cervical margin is most liable to caries, and here the conservative and preservative qualities of tin make it specially applicable.

"Electrolysis demonstrates to us that no single metal can be decomposed, but when gold and tin are used in the above manner they are united at the line of contact by electrolysis. The surface of both metals is exposed to the fluids of the mouth, and the oxid of tin is deposited on the tin, by reason of the current set up by the gold; thus some atoms of tin are dissolved and firmly attached to the gold, but the tin does not penetrate the gold to any great extent." (Dr. S. B. Palmer.)

This connection of the metals assists in holding the filling in place, but it is more likely to break apart than if it was all gold. After electrolysis has taken place at the junction, it requires a cutting instrument to completely separate the tin and gold.

For filling by hand pressure, use instruments with square ends and sides, medium serrations, and of any form or size which will best reach the cavity.

For filling with the hand mallet, use instruments with medium serrations, and a steady medium blow with a four-ounce mallet; in force of blow we are guided by thickness of tin, size of plugger, and depth of serrations, strength of cavity-walls and margins, the same as in using gold. The majority of medium serrated hand mallet pluggers will work well on No. 10 tin of one, two, or three thicknesses. If the tin shows any tendency to slide, use a more deeply serrated plugger. The electro-magnetic, and mechanical (engine) mallet do not seem to work tin as well as the hand mallet or hand force, as the tendency of such numerous and rapid blows is to chop up the tin and prevent the making of a solid mass, and also injure the receiving surface of the filling. In using any kind of force, *always* aim to carry the material to place before delivering the pressure, or blow.

In order to obtain the best results, there must be absolute dryness, and care must be exercised, not thinking that because it is *tin* it will be all right. Skill is required to make good tin fillings, as well as when making good gold fillings. Always use tapes narrower than the orifice of the cavity; they are preferable to rolls or ropes. After a few trials it is thought that every one will have the same opinion. A roll or rope necessarily contains a large number of spaces, wrinkles, or irregularities, which must be obliterated by using force in order to produce a solid filling; thus more force is employed, and more time occupied in condensing a rope, than a flat tape; the individual blow in one case may not be heavier than in the other, but the rope has to be struck more blows. The idea that a rope could be fed into a cavity with a plugger faster and easier than a tape has long ago been disproved. Many of the old-fashioned non-cohesive gold foil operators used flat tapes, as did also Dr. Varney, one of the kings of modern cohesive gold operators.

The tape is made by folding any portion of a sheet of foil upon itself until a certain width and thickness is obtained. This tape is very desirable in small or proximal cavities where a roll or rope would catch on the margin and partially conceal the view.

In the form of a tape, perhaps more foil can be put in a cavity, and there may be more uniform density than when ropes are used. Tapes can also be made by folding part of a sheet of foil over a

thin, narrow strip of metal. Fold the tin into tapes
of different lengths, widths, and thicknesses, according to the size of the cavity; then fold the end
of the tape once or twice upon itself, place it at the
base of any proximal cavity, and begin to condense with a foot plugger of suitable size, and if
there is a pit, groove, or undercut which it does
not reach, then use an additional plugger of some
other form to carry the tin to place; fold the tape
back and forth across the cavity, proceeding as for
cohesive gold. In small proximal cavities a very
narrow tape of No. 10, one thickness, can be used
successfully. For cavities in the occlusal surface,
use a tape as just described, generally beginning at
the bottom or distal side, but the filling can be
started at any convenient place, and with more ease
than when using cohesive gold. In any case if the
tin has a tendency to move when starting a filling,
"Ambler's left-hand assistant" is used, by slipping
the ring over the second finger of the left hand,
letting the point rest on the tin. This instrument
is especially valuable in starting cohesive gold (see
Fig. 6). This is the easiest, quickest, and best
manner of making a good filling, relying upon the
welding or cohesive properties of the tin.

Many operators have not tried to unite the tin
and make a solid mass; they seem to think that it

cannot be accomplished, but with proper pluggers
and manipulation it can be done successfully.

For large occlusal or proximal cavities, the
tapes may be folded into mats, or rolled into cylinders, and used on the plan of wedging or interdigitation, and good fillings can be produced by this
method, but the advantage of cohesion is not obtained, and more force is required for condensing.
They are, therefore, not so desirable as tapes, especially for frail teeth. When using mats or cylin-

FIG. 6.

ders, the general form of the cavity must be depended upon to hold the filling in place. To make
the most pliable cylinders, cut a strip of any desired
width from a sheet of foil and roll it on a triangular
broach, cutting it off at proper times, to make the
cylinders of different sizes.

A cylinder roller, designed by the author, is
much superior to a broach. (See Fig. 7.) When
the cavity is full, go over the tin with a mallet or
hand burnisher, being careful not to injure the cav-

ity-margin. Cut down occlusal fillings with burs or carborundum wheels, and proximal fillings with sharp instruments, emery strips or disks. After partially finishing, give the filling another condensing with the burnisher, then a final trimming and moderate burnishing; by this method a hard, smooth surface is obtained.

Fillings on occlusal surfaces can be faced with No. 20 or 30 tin, and burnished or condensed, by using a burnisher in the engine, but do not rely upon the burnisher to make a good filling out of a poor one.

FIG. 7.

By trimming fillings before they get wet, any defects can be remedied by cutting them out; then with a thin tape (one or two layers of No. 10) and serrated plugger proceed with hand or mallet force to repair the same as with cohesive gold.

Another method of preparing tin for fillings is to make a flat, round sand mold; then melt chemically pure tin in a clean ladle and pour it into the mold; put this form on a lathe, and with a sharp chisel turn off thick or thin shavings, which will be found very tough and cohesive when freshly cut, but they do not retain their cohesive properties for any great length of time,—perhaps ten or twenty days, if kept in a tightly corked bottle. After more or less exposure to the air they become oxidized and do not work well, but when they are very thin they are soft, pliable, and cohesive as gold, and any size or form of filling can be made with them.

Among the uses of tin in the teeth, the writer notes the following from Dr. Herbst, of Germany: "After amputating the coronal portion of the pulp, burnish a mat of tin foil into the pulp-cavity, thus creating an absolutely air-tight covering to the root-canal containing the remainder of the pulp: this is the best material for the purpose." There has been a great deal said about this method, pro and con, notably the latter. The writer has had no practical experience with it, and it need not be understood that he indorses it.

If a pulp ever does die under tin, perhaps it will not decompose as rapidly as it otherwise would, owing to its being charged with tin-salts.

The Herbst method of filling consists in introducing and condensing tin in cavities by means of smooth, highly tempered steel engine or hand burnishers. In the engine set of instruments there is one oval end inverted cone-shaped, one pear-shaped, and one bud-shaped. The revolving burnisher is held firmly against the tin, a few seconds in a place, and moved around, especially along the margins, not running the engine too fast. Complicated cavities are converted into simple ones by using a matrix, and proximal cavities in bicuspids and molars are entered from the occlusal surface. The tin foil is cut into strips, and then made into ropes, which are cut into pieces of different lengths; the first piece must be large enough so that when it is condensed it will lie firmly in the cavity without being held; thus a piece at a time is added until the cavity is full. The hand set of burnishers has four which are pear-shaped and vary in size, and one which is rather small and roof-shaped. In filling and condensing they are rotated in the hand one-half or three-quarters of a turn.

Dr. Herbst claims a better adaptation to the walls of the cavity than by any other method. Proximal cavities in bicuspids and molars can easily be filled; the tin can be perfectly adapted against thin walls of enamel without fracturing them; less annoyance to the patient and less work for the dentist; can be done in half the time required for other methods.

Fees should be reasonably large, certainly more

than for amalgam, for we can save many teeth for a longer time than they could have been preserved with cohesive gold. Many are not able to pay for gold, but they want their teeth filled and *saved*, and it is expected that we will do it properly and with the right kind of material; thus it is our duty in such cases to use more tin and less amalgam.

We should always take into consideration the amount of good accomplished for the patient,—the salvation of the tooth,—and if we are sure, from experience and observation, that the tin filling will last as long as a gold one in the same cavity, or longer, then the fee should be as much as for gold, with the cost of the gold deducted. The amount of the fee ought to be based upon the degree of intelligence, learning, and skill required; upon the amount of nervous energy expended; upon the draft made on the dentist's vitality; upon what benefit has been given the patient; upon the perfection of the result; and, everything else being equal, upon the time occupied; the value of this last factor being estimated in proportion to the shortness of it.

6

CHAPTER VIII.

DR. ROBINSON'S Fibrous and Textile Metallic Filling is a shredded metallic alloy, mostly tin, and has the appearance of woven or felt foil. It is prepared in a machine invented by the doctor especially for the purpose, and he gives directions for using as follows: "Cut the material into strips running with the selvage, and fill as you would with soft foil; use it in all surrounding walls, and finish with a mallet burnisher. Where the surface comes to hard wear, weld on gold with long, sharp serrated pluggers, and finish the same as with gold fillings. The advantage over gold for cervical, buccal, and lingual walls is the perfect ease with which it is adapted, and it can be burnished so as to be absolutely impervious to moisture. Sharp, coarse-serrated pluggers are particularly desirable when using hand pressure." It comes in one-half-ounce boxes, filled with sheets less than two inches square; the thin ones are used for filling, and the thick ones make good linings for vulcanite.

This material is easy to manipulate, but great care is required in condensing at cavity-margins, so as to make a tight filling, and also not injure the margins. It makes as hard a surface as tin foil, and can be cut, polished, and burnished so that it is smooth and looks well; it can be used in temporary or chalky teeth, as a small amount of force condenses it. By using a matrix proximal cavities can be filled from one-fourth to one-half full, and the rest filled with gold, relying on the form of the cavity to hold the gold, regardless of its connection with the fibrous material. If the surface is not overmalleted so as to make it brittle or powdery, a strip of No. 4 cohesive gold, of four or five thicknesses, may be driven into it with a hand mallet and plugger of medium serrations; this union is largely mechanical, but of sufficient tenacity to make manipulation easy, as the material makes a solid foundation to build upon. After exposure to the oral fluids, electrolysis takes place at the junction of the metals.

In 1884 Dr. Brophy said, "I have used Robinson's material for two years, and find it possesses good qualities, and can be used in deciduous teeth, first permanent molars, and cervical margins with better results than can be obtained with any other material by the majority of operators."

Malleted with deeply serrated pluggers, it will make a filling which will not leak. It has saved many teeth from caries at the cervical margin where it might have recurred sooner had cohesive gold been used. In the mouth it changes color about the same as tin foil, and a few fillings did not maintain their integrity, but became crumbly and granular.

For conducting properties it ranks about with tin, and fillings can be made more rapidly than with cohesive gold. We have used ounces of it, but time has proved that everything that can be done with it in filling teeth can also be accomplished as well and in some cases better with tin foil.

In 1878 Dr. N. B. Slayton patented his Felt Foil, which was said to be tin cut into hair-like fibers by a machine, then pressed into small sheets and sold in one-half-ounce books, but it sold only to a very limited extent. Soon after this Dr. Jere Robinson, Sr., invented a machine and began the manufacture of a similar article, but he found he was infringing on the Slayton patent, so he purchased the Slayton machine and made satisfactory terms to continue his own maunufacture of fibrous material. After this little was heard of Slayton's Felt Foil, but Robinson's was considerably used. The two materials look and are manipulated almost exactly alike. Dr. Robinson has both of above-mentioned machines now in his possession.

Archibald McBride, of Pittsburg, Pa., in 1838, made a roll of a portion of a sheet of tin, and then

used just enough gold to cover it, aiming to keep
the gold on the surface, so as to have the filling
look like one of all gold, and not with the idea of
deriving any special benefit from the effects of wear
or preservation as obtained by thus combining the
two metals. The fee for a gold filling was one
dollar; tin, fifty cents. Some operators have advocated using tin and gold (symbol Tg), rolled or
folded together in alternate layers, thus exposing
both metals to the fluids of the mouth; claiming
that fillings can be made quicker, are not so subject
to thermal changes, and can be inserted nearer the
pulp than when gold is used. This may be true
in comparison with gold, but these three claims are
entirely met by using tin alone. Others say that
this union of gold and tin will preserve the teeth as
well as a correct gold filling, making no conditions
or restrictions as to tooth-structure or location of
cavity. They say that it preserves the cervical
margin better than gold; that it expands *slightly*.

A description of some different methods of combining and manipulating tin and gold is subjoined:

(a) Two sheets of No. 4 cohesive gold and one
of the same number of tin are used; place the tin
between the gold, cut off strips, and use with hand
or mallet force the same as cohesive gold; if non-
cohesive gold is used, the strips can be folded into

mats or rolled into cylinders, and are used on the wedging plan, the same as non-cohesive gold, or the strips can be folded back and forth in the cavity until it is full.

(b) Lay a sheet of non-cohesive gold, No. 3, on a sheet of tin of the same number, cut off strips, roll into ropes and use as non-cohesive gold. It is easily packed and harder than tin, and has a preservative action on the teeth. Line the cavity with chloro-balsam as an insulator against possible currents and moisture; especially should this be done in large cavities or chalky teeth.

(c) A sheet of non-cohesive gold, No. 4, is laid on a sheet of tin of the same number, cut into strips and rolled into cylinders, or folded into blocks, always in equal portions; then they will unite to the extent of two leaves. These fillings sometimes become a solid mass about the color of amalgam, and last very well, as the metals have become united by electrolysis. An excess of tin will be marked by lines or pits in the filling, showing where the tin has been disintegrated or dissolved by the chemical action which occurs on the surface exposed to moisture.

No doubt, good fillings have been made by the above methods, yet some were granular, gritty, and were easily removed, while others were quite smooth and hard; probably in the first instance the proportion of tin and gold was not proper,—that is, not equal; or it was not well condensed. Tin being the positive element, it is more easily acted on and disintegrated by electrolysis (chemical action of the fluids).

When this combination does become hard, it wears longer than tin on an occlusal surface, but we believe that in some cases where it was used the teeth could have been saved just as well with either tin or gold, or by filling part of the cavity with tin and the rest with gold.

If tin foil is laid on 22-carat gold and vulcanized, it becomes thoroughly attached and will take a tin polish; the attraction or interchange of atoms takes place to this extent.

This combination of tin and gold can be used at the cervical margin, or a cavity can be lined with it, and the remainder filled with cohesive or non-cohesive gold.

"Tin and gold (Tg) folded or rolled together in equal portions possesses a greater number of desirable properties than any other material, for it is easily adapted, has antiseptic action and a lower conductivity than gold. A new filling is harder than tin, softer than gold, but after a time it becomes as hard as amalgam. It oxidizes and thus helps make tight margins, and is very useful at cervical margins; generally discolors, but not always, and does not discolor the tooth unless a carious portion has been left, and then only discolors that portion. In oral fluids it is indestructible if well condensed, otherwise it is crumbly. There is no change of form, except a *slight* expansion, which does no harm. A weak electric current is set up between the gold and tin, and tin oxid is formed. The hardening and discoloration both depend upon the separation of the tin by the electrical action and its deposition on the surface of the gold. I generally prepare cavities the same as for non-cohesive gold, but a Tg filling may be held in a more shallow cavity and with less undercuts than for gold. Hand pressure is adopted, but a mallet may be used advantageously. Lay a sheet of No. 4 non-cohesive gold on a sheet of No. 4 tin, then cut into strips and twist into ropes; keep the tin on the outside, for it does not tear as easily as gold. Carry the material against the walls and not against the base, otherwise the filling will be flat or concave; but should this occur, then force a wedge-shaped plugger into the center of the filling, and drive the material toward the walls, and then fill the hole or remove all the filling and begin anew.

"In very deep cavities use a mat of Tg, dampened in carbolic acid and dipped in powdered thymol, as a base; this has an antiseptic action, and also prevents pressure on or penetration into the pulp.

"Drs. Abbot, Berlin; Jenkins, Dresden; Sachs, Breslau, have observed tin-gold fillings from fifteen to twenty-five years, and say that for certain cases it is better than any other material. I use square-pointed pluggers (four-cornered), as part of the packing is done with the side of the plugger. Tg is useful in partly erupted molars, buccal cavities under the gums, occlusal cavities in temporary teeth, cavities where all decay cannot be removed. Use Tg with a gold capping in small, deep occlusal cavities, cavities with overhanging walls, occlusal cavities with numerous fissures, large, deep occlusal cavities near the pulp, in proximal cavities.

"Line labial walls of incisors with non-cohesive gold, and fill the remainder with Tg. For repairing gold fillings I use Tg." (Dr. Miller, Berlin, *Dental Cosmos*, 1890.)

Dr. Jenkins, of Dresden, says, "I use Tg in soft, imperfect teeth, of which there are plenty in Germany, because it has pliability, adaptability, slight susceptibility to thermal changes, makes a water-tight joint, very useful at cervical margins, and can be used with a minimum amount of pres-

sure. When packed dry and with the gold next
to the tooth, discoloration occurs only on the surface; packed wet, the whole discolors. I do not
attribute its success to electrical action. Lay a
sheet of No. 4 tin on a sheet of No. 4 non-cohesive
gold, fold so as to keep the gold on the outside; use
the strip with lateral pressure, doubling it upon
itself."

Dr. A. H. Thompson: "After several years'
successful use of tin-gold, I commend it for approximal cavities, cervical margins, and frail walls. The
oxid formed penetrates the enamel and dentin; if a
filling wears down, cover the surface with gold."

Dr. Pearson: "I do not like tin and gold in
alternate layers. I prefer No. 10 tin foil."

Dr. James Truman: "I believe that tin-gold has
a positive value as a filling-material."

"I prepare tin-gold by taking a sheet of No. 4
non- or semi-cohesive gold, fold them together (or
twist them) so as to have the gold on the outside,
and then fill any cavity with it. Since adopting the
above combination I have almost abandoned amalgam. This is recommended on account of its
density, ease of insertion, capacity for fine finish,
non-conducting and non-shrinking qualities, and
compatibility with tooth-substance. Those who
have not used it will be surprised at the rapidity

with which it can be manipulated. It may be employed in any cavity not exposed to view, also in crown, buccal, and approximal fillings which extend into the occlusal surface, as it offers an astonishing resistance to wear. It can be used anywhere that amalgam can, and with more certainty of non-leakage, and it has the additional advantage that it can be finished at the same sitting. Care is necessary in manipulating it, so as to avoid chopping. I use hand pressure when filling, and the mallet to condense the surface." (Dr. A. W. Harlan, *Independent Practitioner*, 1884.)

"Pure tin foil is employed in connection with non-cohesive gold in filling proximal cavities in bicuspids and molars; a sheet of gold and a sheet of tin are folded together." (C. J. Essig: "Prosthetic Dentistry.")

Dr. Benj. Lord says, "A combination in which I find great interest is in the use of soft or non-cohesive gold with tin foil. This is no novelty in practice, but I think that, for the most part, too great a proportion of tin has been used, and hence has arisen the objection that the tin dissolved in some mouths. I am satisfied that I myself until recently employed more tin than was well. I now use from one-tenth to one-twelfth as much tin as gold, and no disintegration or dissolving away of the tin ever occurs. I fold the two metals together in the usual way of folding gold to form strips, the tin being placed inside the gold. The addition of the tin makes the gold tougher, so that it works more like tin foil. The packing can be done with more ease and certainty; the filling, with the same effort, will be harder, and the edges or margins are stronger and more perfect.

"The two metals should be thoroughly incorporated by manipulation. Then, after a time, there will be more or less of an amalgamation. By using about a sixteenth of tin, the color of the gold is so neutralized that the filling is far less conspicuous than when it is all gold, and I very often use such a proportion of tin in cavities on the labial surfaces of the front teeth.

"If too much tin is employed in such cases, there will be some discoloration of the surface of the fillings; but in the proportion that I have named no discoloration occurs, and the surface of the filling will be an improvement on gold in color."

"Dr. Howe. I would like to ask Dr. Lord whether, in referring to the proportions of tin and gold, he means them to be considered by weight?

"Dr. Lord. No, not by weight, but by the width of the strip of tin and the width of the strip of gold. I get the proportions in that way, then lay the tin on the gold and fold the gold over and over, which keeps the tin inside the gold.

"Dr. Howe. Will Dr. Lord tell us whether he refers to the same numbers of gold foil and tin foil; as, for instance, No. 4 gold and No. 4 tin?

"Dr. Lord. I use the No. 5 gold, and tin, I think, of about the same number, but I always use No. 5 gold, both cohesive and non-cohesive."— *New York Odontological Society Proceedings*, 1893, page 103.

"Tin and gold, in the proportions generally used, do not present a pleasing color; when finished, it looks but little better than tin, and after a short time it grows dark, and sometimes black. I use five parts of gold to one of tin, prepared as follows: Lay down one sheet of Abbey's non-cohesive gold foil, No. 6; upon this place a sheet of No. 4; upon this place a sheet of White's globe tin foil, No. 4; upon this another sheet of Abbey's non-cohesive gold, No. 4; upon this a sheet of No. 6. Cut into five strips and crimp; the crimped strips are cut into pieces a little longer than the depth of the cavity to be filled; some of the strips are rolled into cylinders, others are left open, because easier to use in starting a filling. The color of this combination is slightly less yellow than pure gold, and hardens just as rapidly as when the proportions are one to one, but does not become quite as hard. This preparation is non-cohesive, and should be inserted by the wedge process. I use it in the grinding surface of molars and bicuspids, buccal cavities in molars and bicuspids, cervical fissure pits in superior incisors, proximal cavities in bicuspids and molars. If proximal cavities are opened from the occlusal surface, the last portion of the filling should be of cohesive gold to withstand mastication. In simple cavities I place as many pieces as can be easily introduced, using my pliers as the wedging instrument to make room for the last pieces, and then condense the whole. If the cavity is too deep for this, I use Fletcher's artificial dentin as a base, because it partly fills the cavity and the ends of the cylinders stick to it. After an approximal cavity is prepared, use a matrix held in place by wooden wedges; the cylinders are about one-eighth of an inch long, and condensed in two or three layers so as to secure perfect adaptation; hand pressure is principally used, but a few firm strokes with a hand mallet are useful. When ready to add the cohesive gold for the grinding-surface, a few pieces of White's crystal mat gold should first be used, because it adheres beautifully; thus a perfect union is secured, but I never risk adding the gold without leaving a little undercut for it in the tooth. By this method we obtain a beautiful contour filling in a short time. Fillings should be burnished and then polished with a fine strip, or moistened pumice on a linen tape. Where cohesive gold is used for the entire filling, in many cases the enamel-walls, already thin near the cervical margin, are made thinner by the unavoidable friction of the polishing strips, but tin and gold is so soft that a good surface is obtained in a few moments, and this danger is reduced to a minimum. The surface is as smooth as a cohesive gold filling, while such a surface is impossible with non-cohesive gold. In cavities which extend so far beyond the margin of the gum that it is impossible to adjust the rubber-dam, I prepare the cavity as usual, then adjust a matrix, disinfect, dry, and fill one-third full with tin and gold, then remove the matrix, apply the rubber, place matrix again in position, and complete the filling by adding a little tin and gold, then pure gold." (Dr. W. A. Spring, *Dental Review*, February, 1896.)

Dr. T. D. Shumway says, "To have a scientific method of treatment, there certainly must be a recognition of what is known of the nature of tooth-structure. The method adopted more than a quarter of a century ago, and which is at present employed, does not accord with the teachings of the physiologist and microscopist; it is in direct opposition to natural law. Each new discovery in the minute structure of the teeth makes this more plain; pounding the teeth with a mallet cannot be defended on scientific grounds. That it has not resulted more disastrously is due to the wonderful recuperative energy of nature to repair injury. No one would think of attempting to arrest and prevent disintegration in any other vital organ by abrasion. Why, then, in operation on the teeth, should we reverse the plain, simple teaching of nature? Placing cohesive gold against the dentinal walls by pounding it to heal a lesion is opposed to natural law. Cohesive gold will not be mastered by force; if compelled to yield by superior strength, it seeks a way to release itself; it is easily coaxed, but not easily driven. Cohesive gold will unite with tin at an insensible distance just as cohesive gold unites with itself; this union takes place without force or pressure. Exactly what takes place when gold and tin are brought in contact in the way described we do not know; we can only say that there appears to be a perfect union. When cohesive gold was introduced to the profession, while it was softer than non-cohesive foil, it was found to resist under manipulation. This resistance is in accordance with the well-known law that all crystalline bodies, when unobstructed, assume a definite form. With gold the tendency is to a spherical form. The process of crystallization is always from within outward. The mallet was introduced to overcome the resistance caused by the development of the cohesive property. Pounding gold with a mallet only increases its crystallization. A crystalline body coming in contact with a fibrous one can neither be antiseptic nor preservative; a filling-material which possesses these properties must be one that corresponds or is in harmony with tooth-substance.

"In the interglobular spaces there is a substance which is called amorphous or structureless, and a filling to be in harmony with this substance should be amorphous or structureless in its composition. The only materials we have which meet these conditions are gutta-percha and tin. It is its structureless character that gives to tin its value. Coming in contact with the living dentin, it is easily adapted, and does not excite inflammation; it does not interfere with the process going on within the teeth to heal the leison caused by caries. A wound from a bullet made of tin, unless it struck a vital part, nature would heal, even if the cause of the wound was not removed, by encysting the ball. This process of nature of repairing injury by en-

7

cysting the cause is of interest to the dentist in the study of suitable filling-materials. Tin is very useful at the cervical margin of cavities; it acts as an antiseptic or preservative, and reduces the liability to subsequent decay. It is our endeavor to obtain a filling that will preserve the teeth and reduce the liability to, if not wholly prevent, secondary decay. The law of correspondence is of more consequence than the mechanical construction of the filling. Tin can be used without that rigid adherence to mechanical rule that is necessary to retain a filling of cohesive gold; thus less of the tooth needs to be sacrificed.

"Gold will unite with tin under certain conditions so as to form apparently a solid mass. By a combination of these metals, not by interlacing or incorporating one in the other, but by affinity, secured by simple contact, we have all the preservative qualities of tin combined with the indestructible properties of gold. For the base of the filling we have a material in harmony with tooth-substance, introduced in a way that is in accord with the law that governs all living bodies, and for the outside a crystalline substance that corresponds to the covering of the teeth. This covering of gold is a perfect shield to the base, and the field for the display of artistic skill in restoring contours is as broad as though gold was used entirely. Will a filling of this kind withstand mastication? There is in the economy of nature a provision made to overcome the resistance of occlusion. The teeth are cushioned in the jaw and yield under pressure. The elasticity of the substance of which the teeth are made is well understood. Ivory is the most elastic substance known. The teeth coming together is like the percussion of two billiard balls. Now a filling to save the teeth should correspond as nearly as possible with the tooth-substance; it should not be arbitrary, but elastic and yielding. Tin is interdigitous; it expands laterally, and is almost as easily introduced as amalgam, and when put in place does not have to be bound to be retained. Tin, with an outside covering of gold to protect it, makes a filling to which amalgam bears no comparison. In the light of scientific investigation there can be but one method—a method based upon the recognized principle that the *filling-material* and the *manner* of *introducing* it shall correspond to and be in harmony with the living, vital organism with which it comes in contact.

"After excavating, the cavity is treated with absolute alcohol, as cleanliness and thorough dryness are absolutely essential.

"The *tin* is put in with steel pluggers, after the method of wedging; it must be thoroughly condensed, so as to leave a smooth surface, and enough used to come up to where the enamel and dentin join.

"The effect is not produced by incorporating or interlacing the gold with the tin; we rely upon the affinity of the two metals to retain the gold; no undercuts, angles, or pits are made in the tin, dentin, or enamel. The gold, extra cohesive from No. 4 to 40, is made to adhere to the tin by simple contact, without pressure or force; the union is not mechanical.

"The instruments used for filling the remainder of the cavity with gold are Shumway's ivory points, which adapt the gold nicely to the margin.

"The set consists of five and were patented in 1881, and have been used by me since that time for manipulating cohesive gold. One 'point' is for proximal cavities in the anterior teeth; three 'points' of different sizes are for occlusal cavities; one 'point' for proximal cavities in bicuspids and molars and labial and buccal cavities; the sides, edges, and ends of the 'points' are used, as the purpose is simply to obtain contact.

"The 'point' shown full size in Fig. 8 is of more general application than any of the others, and is

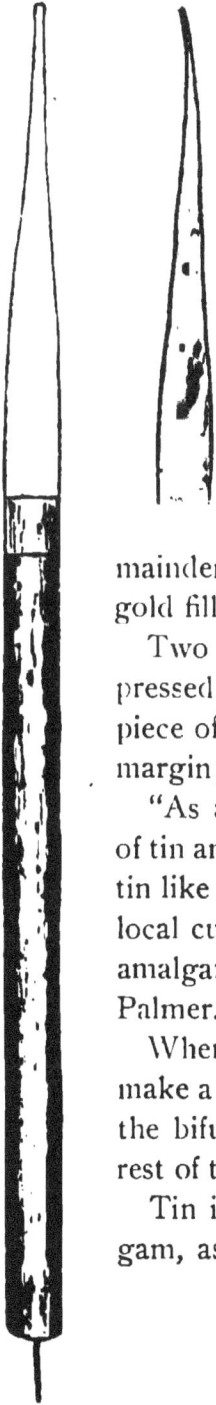

FIG. 8.

used for proximal cavities in bicuspids, also labial and buccal cavities. The handle is made of ebony, and has a silver ferrule, from which the ivory extends to the end and completes the instrument.

"The metal pin in the end of the handle is for picking up and carrying the gold."

Tin has been used successfully for completely lining cavities, filling the remainder with gold; it is also useful for repairing gold fillings.

Two or three thicknesses of tin foil may be pressed into a cavity with a rubber point or hard piece of spunk, allowing it to come well out to the margin; filling the rest with amalgam.

"As a lining it presents to dentin an amalgam of tin and mercury which does not discolor the dentin like ordinary amalgam, and helps do away with local currents on the filling, which is one cause of amalgam shrinkage in the mouth." (Dr. S. B. Palmer.)

When caries extends to the bifurcation of roots, make a mat of two or three layers of tin, place it in the bifurcation and use it as a base in filling the rest of the cavity with amalgam.

Tin is second in importance in alloys for amalgam, as it increases plasticity, prevents discoloration, reduces conductivity and edge strength, retards setting, favors spheroiding, therefore should not be the controlling metal.

It will be noticed that when cavities are lined with tin foil, it only constitutes a small part of the filling, and that it has not been melted with the other metals in the alloy before being amalgamated.

A thick mat of tin has been recommended as a partial non-conductor under amalgam fillings.

Plastic tin can be made by pouring mercury into melted tin, or by mixing the fillings with mercury at ordinary temperatures; it has a whitish color, and if there is not too much mercury it occurs in the form of a brittle granular mass of cubical crystals. Generally amalgams of tin and mercury do not harden sufficiently, but forty-eight parts of mercury and one hundred of tin make a fairly good filling, said to have a therapeutical value; it should not be washed or squeezed before using, and "is not a chemical combination."

"Tin unites with mercury in atomic proportions, forming a weak crystalline compound." (Dr. E. C. Kirk.)

Mercury and tin readily unite as an amalgam under ordinary circumstances, and form a definite chemical compound having the formula Sn_2Hg. (Hodgen.)

Another preparation of tin is known as stannous gold; it is manufactured in heavy sheets and used the same as cohesive gold foil, and can be easily manipulated, for it is rather plastic.

Crystal tin for taking the place of tin foil:

"Take chemically pure hydrochloric acid and dissolve tin foil in it until a saturated solution is obtained; this may be done speedily by heating the acid to a boiling point, or the same thing can be accomplished in a few hours with the acid cold; it is then chlorid of tin. It is then poured into a clean vessel and an equal quantity of distilled water added; then a clean strip of zinc is plunged into the solution, and tin crystals are deposited on the zinc; when there is sufficient thickness on the zinc, remove both, and slip the crystals off from the zinc into pure water, clean the zinc thoroughly, and reinsert for another coating. The character of the crystallization will be modified by the extent of the dilution of the solution in the first place. Wash the tin in pure water until all traces of the acid are removed, or a few drops of ammonia can be added to neutralize the acid. It was suggested that it would be desirable to have some acid remain in the tin for filling teeth in which there is no sensitive dentin. We have put in a few fillings, and it works beautifully, and makes firmer fillings than foil. It must be kept in water (probably alcohol is better). It is pure tin, unites perfectly, and works easier than foil." (Dr. Taft, *Dental Register of the West.* 1859.)

For some years it was considered the best practice to enlarge all root-canals and fill them with gold; in many of these cases the crown cavities were filled with tin.

Tin has been used for filling root-canals, but should there happen to be any leakage through the foramen or tooth-structure, the tin will discolor, and there may be infiltration into the crown. thus causing discoloration, which might be objectionable if the crown was filled with gold. Chloropercha, gutta-percha, and oxychlorid of zinc are much better for this purpose.

The apical quarter of a canal has been filled with tin, and the remainder with cement. Tin can be used for filling root-canals. Roll on a broach small triangular pieces of the foil into very small cone-shaped cylinders, carry to place, then withdraw the broach, and force in the cylinder with the same or a larger broach; sometimes it is necessary to use another broach, to push the cylinder off from the one on which it is rolled. Another method is to carry and pack into the canal by means of a broach, very narrow strips of No. 10 or 20 foil; or the apical third of the canal could be filled with gold and the rest with tin.

"About four years ago I concluded to try tin for filling root-canals; then I began to look for patients whose general health was good, who had strong, hardy-looking teeth, and kept their mouths in good condition. I found one who answered all my requirements, with a molar to be filled, and they would not have it filled with gold, or could not, on account of the expense. I filled the canals with tin and the crown with amalgam. After filling thirty-eight molars in this way I stopped for developments. In six or seven weeks a lady returned with an inferior molar abscessed, but at the time it was filled the circumstances were such that it could not be properly treated. In nine months a gentleman for whom I had filled four molars returned with an inferior one abscessed. This is the sum-total of abscessed teeth where tin was used in the root-canals, at the end of four years. The others are in good condition, as I have seen them every six months. The roots were treated from four to six weeks with carbolic acid before filling." (Dr. A. W. Harlan, *Missouri Dental Journal*, 1872.)

"Tin foil is just as good as gold for filling root-canals, as it is entirely innocuous and sufficiently indestructible, while its softness and pliability commend it. Where gold is to be used for the crown, it is better to fill the bulbous portion of the pulp-cavity with gold also, so as to weld these portions of gold together. The success of Dr. Harlan's treatment was about equal to what might be expected from the same number of teeth where the canals had been filled with gold." (Editor *Missouri Journal.*)

Shavings turned from a disk of pure tin have been used in combination with Watt's sponge gold for filling teeth, either by making a portion of the filling from each metal or using them indiscriminately.

A mat of tin foil dipped in chloro-percha can be used to cap an exposed pulp, or a concave tin disk can be used for the same purpose. A mat of tin has been used over a slight exposure of the pulp, because of its slight conduction of heat and cold, thus avoiding much thermal irritation and stimulating recuperation.

Some use Robinson's fibrous material as a surface for tin fillings, thinking that it is harder and will wear longer because of the erroneous notion that it has platinum in it.

CHAPTER IX.

TIN has been recommended for temporary fillings in sensitive cavities, because it is soft and easily packed in contact with the walls, has therapeutic value, and after a time, when the temporary filling is removed, the cavity is not as sensitive as formerly.

It has been observed that starting gold in a sensitive cavity causes pain, but starting tin in the same place seldom does.

As long as tin preserves its integrity it preserves the tooth, therefore tin fillings should not be repaired with amalgam, as their integrity may be destroyed. Cavities can be partly filled with tin and completed with sponge, fibrous, or crystalloid gold, after the manner described for beginning with tin and finishing with gold foil.

"I advocated tin at the cervical wall, cervico-lingual and cervico-buccal angles to the thickness of 24 plate. Then complete the filling with gold. Some of my most successful efforts in saving soft teeth have been made in this way. This method has great value over gold for the whole filling, but there are two objections to it: First, it imparts to the cervical border the color and appearance of decay, so that in three cases where an instrument passed readily into the tin I have removed the fillings, without any necessity for it, not even finding any softening of the margins. Second, its use requires the same conditions of dryness, shape of cavity, delicate manipulation, inconvenience to patient, and strain upon the operator as when gold is used alone." (Dr. D. D. Smith, *Dental Cosmos*, 1883.) He admits that this method saves *soft* teeth and also cervical margins. Do not those two very important factors more than counterbalance the color, and oversight of the dentist?

Dryness is an essential in making the best filling with any material, and the time and strain consumed by the majority of operators in filling with tin is not more than one-half what it is in using gold.

"I use tin at the cervical margin of all proximal cavities in bicuspids and molars. I prepare a matrix of orange-wood to suit each case, letting it cover about one-third of the cavity, then fill with tin condensed by hand force and automatic mallet; now split the matrix and carefully remove it piece by piece, so as not to disturb the tin; then trim and finish this part of the filling. Make another wooden matrix, which covers the tin and remainder of the cavity, and fit it snugly to place. Use a coarsely serrated plugger and begin packing non-cohesive gold into the tin, letting it fill about one-third more of the cavity; then complete the last third (surface) with cohesive gold. I have tested this method for twenty years, and it has given me splendid results. I always tell patients that there will appear sooner or later a slight discoloration near the gum, which must not be mistaken for caries." (Dr. A. P. Burkhart.)

Another use for tin in the operating-room is found in Screven's "Gutta-percha-coated Tin Foil," a cohesive, antiseptic non-conductor, of which the inventor says: "Cement fillings that have been kept dry for ten hours after mixing will be much harder than those soon exposed to moisture, and they will retain that hardness though exposed to moisture afterward. This preparation will keep a filling perfectly dry in the mouth, and when removed the filling will be found hard as stone. There is nothing better for lining cavities, holding nerve-caps in position, holding a preparation in place when devitalizing a pulp where the tooth is so much broken away as to make it difficult to prevent a filling showing through the enamel, and for many other purposes."

High-heat gutta-percha has been used as a base in deep occlusal, buccal, and approximal cavities, completing the fillings with tin. Occlusal cavities may be filled with tin; then after the filling is condensed and finished, drill out the center and fill with cohesive gold, not cutting away the tin so as to expose the margin; such fillings wear well, as much of the attritial force comes on the gold portion of the filling.

With the exception of the part in brackets, the following article is from the *British Journal*, May, 1887:

"If a person eats an oyster stew at 130° F., a gold filling would carry the difference between the temperature of the stew and that of the mouth, $130 - 98 = 32°$, almost undiminished to the bottom of the cavity; allowing 2° of diminution, then the cavity around the gold filling has assumed 128°; now the person feels warm and drinks ice-water at 32°. Taking into consideration the specific heat of the gold filling, it will assume about 40°, which it carries with a diminution of the cold of about 4°,—that is, as if it was 44°,—into the interior of the cavity; then the cavity will assume 44°, the difference within one-tenth of a minute being $128 - 44 = 84°$, a change which would produce a violent inflammation in any organ which was not accustomed to it. This derangement in the tooth means interruption of circulation, and young teeth will be most affected.

"Thermal effect depends on heat-conducting power [gold is nearly four times as good a conductor of heat as tin] and also on specific heat, so the more the latter approaches that of the tooth the less it is liable to produce sudden changes [thus favoring tin]. Specific heat manifests itself by the speed of changes, while the heat-conducting power influences the intensity [then the intensity of heat in a gold filling would be three or four times as much as in a tin filling]. In speed gold produces this change in one-tenth of a minute" [tin in one-fifth,—that is, gold absorbs heat and expands about twice as fast as tin].

In 1838 Dr. J. D. White introduced sharp-wedge-shaped instruments for filling teeth, and he claims to have been the first to use them; they pack laterally as well as downward, and present as small a surface to the filling as possible, so that the greatest effect may be produced upon a given surface with a given power. Rolls of either tin or gold are made by cutting any desirable portion from a sheet of No. 4 foil; cut this portion once transversely, place on a napkin or piece of chamois, then with a spatula fold a very narrow portion of the edge once upon itself; then with the spatula resting on the thickened edge draw the spatula away from it with gentle pressure, and the foil will follow in a roll.

The old method of using rolls, ropes, and tapes or strips is the same, but we will describe one method of using tapes. (See Fig. 9.) A *strip* is a single thickness of foil in ribbon form; a strip.

FIG. 9.

folded lengthwise once, twice, or more forms a *tape* of two, four, or more thicknesses of foil. The tin foil should be cut into strips and folded into tapes proportioned in width and thickness to the size of the cavity. One end of the tape is carried to the bottom of the cavity and then forced against the side opposite the point where we intend to finish; now remove the wedge-shaped plugger and catch the tape outside of the cavity, and fold another portion against that already introduced, letting all the folds extend from the bottom to a little beyond the margin. Proceed in this manner, with care and sufficient force, until the cavity is full, using for the last folds a small instrument. Condense the surface with a large plugger, then go over it carefully with a small instrument, and if any part yields, force in a wedge-shaped plugger and fill the opening in above-described manner; condense, burnish, and trim alternately until the surface is level with the cavity-margin. By extending the folds from the orifice to the base of the cavity, the liability of the tin to crumble or come out is effectually prevented, and by putting it in with a wedge-shaped plugger it is pressed out into all depressions of the walls.

A later method of filling with tape or rope is to use wedge-shaped pluggers with sharp serrations, filling the *ends* of the cavity, and as the two parts approach each other that next to the wall should be in advance of the rest, thus an opening will be left in the center which can be filled with a smaller tape or rope.

Another old method: Take a piece of foil and roll it into a hard ball; then gradually work it into

8

the cavity, being careful to have sufficient around
the margin.

Still another suggested method: Roll a piece
of foil into a loose ball, place it in the cavity, and
pass a wedge-shaped plugger into its center.   This
has the effect of spreading the tin toward the walls
of the cavity, the opening to be filled with folds in
a way already described.   The wedge is used as

FIG. 10.

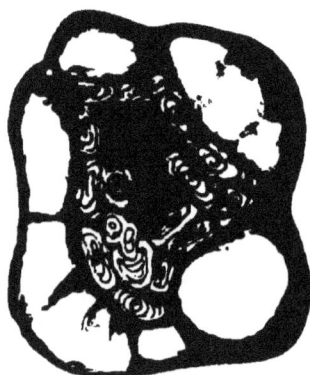

often as it can be made to enter, filling each opening with folds; then condense the surface, trim, and
burnish.

The English give the Americans the credit of
first using cylinders.   Anyhow, Dr. Clark, of New
Orleans, in 1855, used them made from non-
cohesive gold, and also from gold and tin in alternate layers.   (See Fig. 10.)

Cylinders were used which were a little longer than the depth of the cavity, introduced with wedge-shaped pluggers around the walls, each one being closely adapted to the margin; then another row was added, which was forced firmly against the preceding, continuing this process until the cavity was full. The wedge, having a smooth end and sides, is forced into the center so as to drive the tin toward the sides of the cavity, being careful not to split the tooth; the opening is then filled with a cylinder. Now force a smaller-sized wedge into the center of the last cylinder, and into the opening introduce another cylinder, proceeding in this manner until the filling is solid. Then condense the ends of the cylinders, trim, and burnish. For the same operation more recent pluggers are wedge-shaped, with sharp, deep serrations. In these cases the filling is retained by the general form of the cavity and wedging within a certain limit, and not by cohesion of the different parts. For a time tin cylinders were prepared and put on sale at the dental depots.

As far as we are aware, the first tin foil made use of in operative technics was by Dr. F. S. Whitslar, who removed a disk of German silver from an ivory knife-handle in 1845, then used hand pressure to fill the cavity with tin. In the college course of operative technics tin foil can be used, almost to the exclusion of gold foil, to demonstrate the manipulation of both cohesive and non-cohesive gold. Shavings scraped from a bar of tin are also useful in operative technics; they are more cohesive than foil.

www.ingramcontent.com/pod-product-compliance
Lightning Source LLC
Chambersburg PA
CBHW021826190326
41518CB00007B/762